S104 Exploring science

Science: Level 1

The

Quarks to Quasars

Prepared by Andrew Norton

This publication forms part of an Open University course S104 *Exploring science*. The complete list of texts which make up this course can be found on the back cover. Details of this and other Open University courses can be obtained from the Student Registration and Enquiry Service, The Open University, PO Box 197, Milton Keynes MK7 6BJ, United Kingdom: tel. +44 (0)845 300 60 90, email general-enquiries@open.ac.uk

Alternatively, you may visit the Open University website at http://www.open.ac.uk where you can learn more about the wide range of courses and packs offered at all levels by The Open University.

To purchase a selection of Open University course materials visit http://www.ouw.co.uk, or contact Open University Worldwide, Michael Young Building, Walton Hall, Milton Keynes MK7 6AA, United Kingdom for a brochure. tel. +44 (0)1908 858793; fax +44 (0)1908 858787; email ouw-customer-services@open.ac.uk

The Open University
Walton Hall, Milton Keynes
MK7 6AA

First published 2008

Copyright © 2008 The Open University

All rights reserved. No part of this publication may be reproduced, stored in a retrieval system, transmitted or utilised in any form or by any means, electronic, mechanical, photocopying, recording or otherwise, without written permission from the publisher or a licence from the Copyright Licensing Agency Ltd. Details of such licences (for reprographic reproduction) may be obtained from the Copyright Licensing Agency Ltd, Saffron House, 6–10 Kirby Street, London EC1N 8TS; website http://www.cla.co.uk/.

Open University course materials may also be made available in electronic formats for use by students of the University. All rights, including copyright and related rights and database rights, in electronic course materials and their contents are owned by or licensed to The Open University, or otherwise used by The Open University as permitted by applicable law.

In using electronic course materials and their contents you agree that your use will be solely for the purposes of following an Open University course of study or otherwise as licensed by The Open University or its assigns.

Except as permitted above you undertake not to copy, store in any medium (including electronic storage or use in a website), distribute, transmit or retransmit, broadcast, modify or show in public such electronic materials in whole or in part without the prior written consent of The Open University or in accordance with the Copyright, Designs and Patents Act 1988.

Edited and designed by The Open University.

Typeset by SR Nova Pvt. Ltd, Bangalore, India.

Printed and bound in the United Kingdom by University Press, Cambridge.

ISBN 978 0 7492 2672 5

1.1

Contents

Chapter 1 Introduction – the Universe large and small

> The most incomprehensible thing about the Universe is that it is comprehensible.
>
> Albert Einstein, German-Swiss-American physicist (1879–1955)

In Book 6, you saw how the age of the Earth can be measured and how events in its past history may be studied by looking at the record laid down in the Earth's rocks. In this book, the scope of the course is widened in both time and space. In fact, you will be concerned with the *entire* history of the Universe from its origin to the present day. You will also be going way beyond the everyday scale of familiar Earth-bound objects and processes to examine both the smallest and the most distant structures that make up the Universe. It is convenient therefore to begin by looking at the various sizes of things, and the various distances to objects, that comprise the Universe. In scientific terms, these sizes and distances are jointly referred to as **length scales**.

The words 'quark' and 'quasar' may be less than a page apart in a scientific dictionary, yet the objects they refer to are as far apart as it is possible to be in terms of the length scales of the Universe. **Quarks** are the fundamental constituents of matter and are smaller than 10^{-19} m in size, whilst **quasars** represent the most distant astronomical objects it is possible to observe and are up to 10^{26} m away.

■ How much bigger is the distance to a quasar than the size of a quark?

□ $\dfrac{10^{26}\ \text{m}}{10^{-19}\ \text{m}} = 10^{26+19} = 10^{45}$.

So the distance to a quasar is 45 orders of magnitude greater than the size of a quark. (If you need to remind yourself how to combine powers of ten such as this, refer to Book 2, Box 13.1.)

These two length scales – separated by a factor of a billion, billion, billion, billion, billion – represent the extremes of human comprehension of the Universe. Quarks and quasars therefore serve as convenient limits between which we might attempt to understand the Universe as a whole.

Before beginning this exploration, it is worth pausing for a moment to try to appreciate the sheer range of length scales implied by the simple phrase 'quarks to quasars'. A factor of a billion (10^9) is nine orders of magnitude. In terms of sizes, this is equivalent to going from, say, the size of a marble (about 1 cm, or 10^{-2} m, in diameter) to the size of the planet Earth (to the nearest order of magnitude, 10 000 km, or 10^7 m, in diameter). But there are *five* such steps in going from quarks to quasars, *each* requiring an increase in size by a factor of a billion! Table 1.1 and Figure 1.1 summarise the 45 orders of magnitude in this journey from one extreme of the Universe to the other.

Table 1.1 Five factors of a billion on the journey from quarks to quasars.

Size scale	Item	Comment
10^{-19} m	Quarks	the fundamental constituents of the Universe
… are a billion times smaller than …		
10^{-10} m	Atoms	the building blocks of matter
… are a billion times smaller than …		
10^{-1} m	Apples	a typical size scale in the everyday world is 10 cm
… are a billion times smaller than …		
10^{8} m	Jupiter	giant planets are of order 100 000 km in diameter
… is a billion times smaller than the distance to …		
10^{17} m	nearest stars	the nearest stars to the Sun are within 10 light-years
… are a billion times nearer than …		
10^{26} m	Quasars	the most distant objects we can observe are so far away, the light emitted by them has taken billions of years to reach us

Understanding how the Universe works throughout this range of length scales is necessarily the most wide-ranging subject that can be addressed by science. Other scientific issues, such as global warming or genetic engineering, certainly have more immediate relevance to our everyday lives, but when it comes to fundamental questions such as:

- How does the Universe behave on small and large scales?
- What rules does the Universe follow?
- How does the Universe change with time?

there are simply none that are bigger in scope. Answers to questions like these are to be found in the fields of **cosmology** and **particle physics**. Scientists who work in these two apparently unrelated areas of science – one concerned with the infinitely large, the other with the unimaginably small – have come together in recent years in an attempt to explain the universal processes that occur throughout time and space. Earlier in this course, in Book 3, you learned about the principles of physics that govern processes on an *everyday* scale – in particular, energy conservation and the behaviour of electromagnetic radiation. In contrast, this book explores the physics of the very small and the very large. It will take you on a journey from quarks to quasars and introduce you to the key ideas of modern physics, developed over the last century, that now form the fundamental basis for our understanding of how the Universe works.

This book will provide you with the opportunity to develop and demonstrate many of the learning outcomes for the course. In particular, it includes development

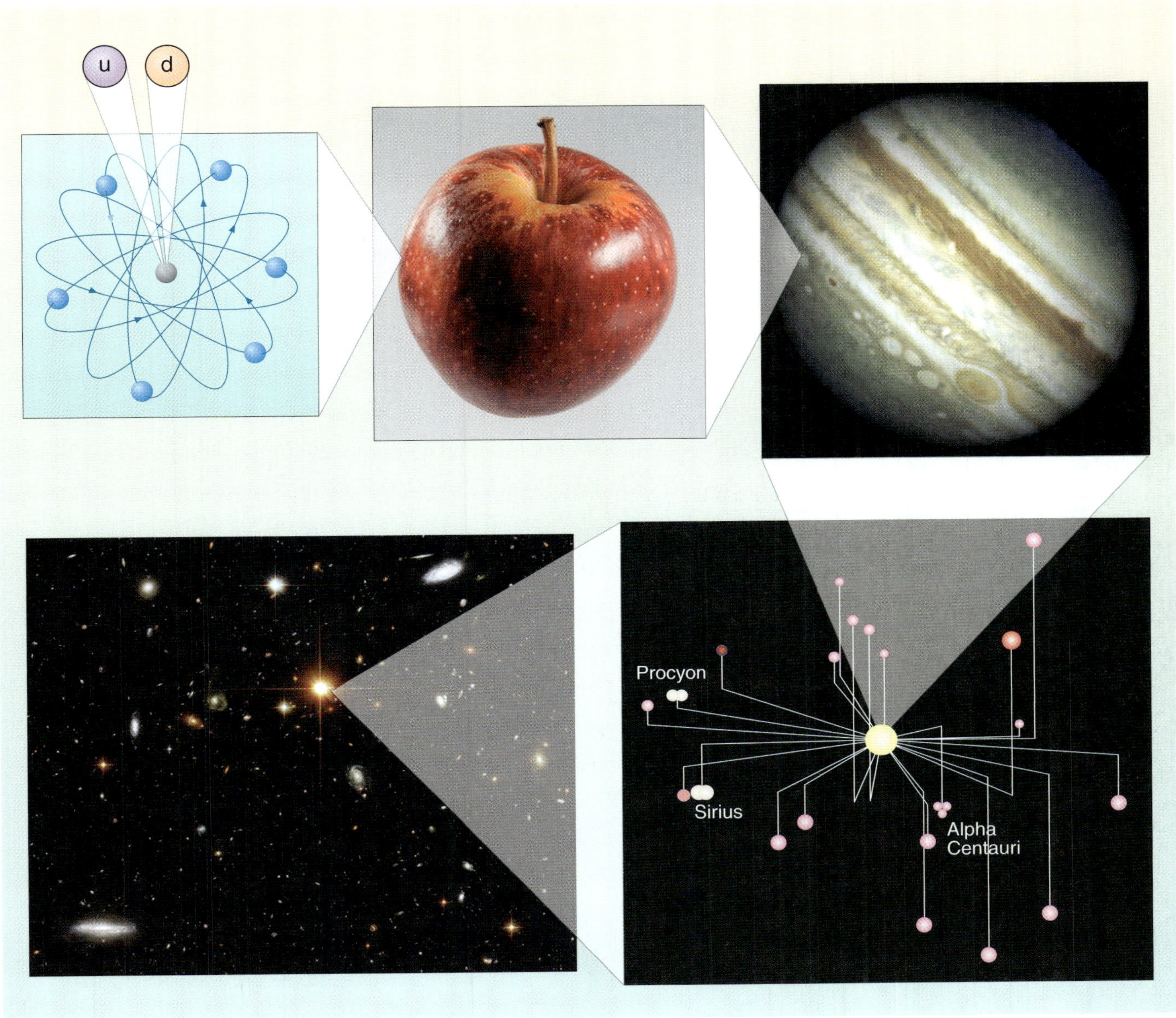

Figure 1.1 Quarks (shown here by the particles with 'u' and 'd' labels; you will meet these in Chapter 7) are a billion times smaller than atoms, which are themselves a billion times smaller than apples. Apples in turn are a billion times smaller than a giant planet like Jupiter. The size of Jupiter is a billion times smaller than the distance to the nearest stars, which are themselves a billion times nearer than quasars.

of the various mathematical skills that have been introduced earlier in the course, as well as development of the scientific language, concepts and principles appropriate to a study of cosmology and particle physics.

1.1 Cosmology and particle physics

Cosmology is the branch of science that involves the study of the Universe as a whole. The research tools of cosmologists include powerful telescopes, such as those in Figure 1.2, that are able to detect galaxies out to the farthest reaches of the Universe. You will get a flavour of this work later by using a computer-based 'virtual telescope' to measure the expansion rate of the Universe, and so calculate its age. It may seem strange that people working in this field should count particle physicists amongst their closest allies. The research tools of particle physicists include giant particle accelerators, such as that shown in Figure 1.3, in which high-energy beams of particles are smashed together, enabling details of exotic reactions to be investigated and understood. But this is the key to the union of these two subjects. For only in particle accelerators are scientists able to recreate the high-energy conditions that once existed in the Universe during the first moments of its creation. When particle physicists study these reactions they can provide cosmologists with a window on the Universe when it was only one-thousandth of a billionth of a second old.

Figure 1.2 The twin domes of the W. M. Keck Observatory, 4200 m above sea level on the summit of Mauna Kea, Hawaii. These massive domes each house a telescope whose primary mirror is 10 m in diameter. When they were completed in 1997 these telescopes were the largest in the world, enabling astronomers to observe objects up to 250 million times fainter, and see detail on a scale 500 times finer, than can be seen with the naked eye.

Figure 1.3 The Large Hadron Collider (LHC) at CERN, the European Organisation for Nuclear Research, the world's largest particle physics centre, near Geneva in Switzerland. The LHC consists of a ring 27 km in circumference, part of which is shown here, within which two high-energy beams of protons will collide head-on. Just completed in 2007, it will enable physicists to recreate conditions that existed only one-thousandth of a billionth of a second after the Universe began.

An example of the interplay between these two areas of study concerns the fundamental particles known as neutrinos, which you will read about in Chapter 7. A few years ago, cosmologists studying reactions that occurred in the early Universe announced that there can be no more than three types of neutrino. If there were, say, four types of neutrino then they calculated that there would be more helium in the Universe than is actually observed. Particle physicists, studying decays of exotic particles in their high-energy accelerators, were also able to calculate how many types of neutrino there are in the Universe. The answer the particle physicists arrived at was also three – if there were more, or fewer, types of neutrino, the particles they were studying would have decayed at a different rate. So we can be fairly confident that there really are only three types of neutrino in the Universe – whether the problem is tackled from the large or the small scale.

The important idea is that cosmology and particle physics complement each other. They are two research areas at the forefront of scientific understanding, and we hope that you will find learning about them to be both stimulating and rewarding.

1.2 Understanding how the Universe works

The big questions outlined at the start of Chapter 1 will now be used as themes for what follows.

In the first part of the book, comprising Chapters 2 to 7, you will be concerned with the overall *structure* and *composition* of matter on the smallest scales. The branch of science that describes how the smallest particles of matter behave is known as **quantum physics**. After reminding you of relevant information from earlier in the course and the questions posed by what you learned there, the two key features of quantum physics that control the behaviour of matter and energy at a fundamental level – quantised energy and indeterminate positions and velocities – are explored. Taking the world apart, the structure of atoms is then examined before moving on to look at how atomic nuclei behave, and finally you will look inside the protons and neutrons to discover quarks – the fundamental building blocks of the Universe. This is quite an itinerary for your journey into the heart of the atom. You will be working in the tiny, subatomic domain and will be coming to terms with ideas that are almost unbelievable in the context of the everyday world. Most people are awe-struck by spectacles such as mountain scenery here on Earth or an exploding star in outer space. Prepare yourself to encounter phenomena that are much too small to be seen but that are none the less just as rich in fascination and mystery.

In the second part of the book, comprising Chapters 8 and 9, you will turn to the science of cosmology and examine the properties of the Universe on the largest scales. You will see that observations of quasars hold the key to understanding the distant Universe – by which is meant distant in both time and space. You will discover that the Universe is not static: it was different in the past to how it is now, and it will be different again in the future. Our understanding of this evolution relies crucially on two pieces of evidence: first, evidence that the Universe is *expanding* and, second, evidence that the Universe is *cooling*. Each of these is considered in turn to complete the picture of the Universe as it is today. Once again, you will be dealing with concepts that lie completely outside those of everyday experience.

Any attempt to chart the history of our evolving Universe must take account of the laws that govern all physical processes. So the third part of the book, comprising Chapters 10 to 13, contains accounts of the distinctive features of the four types of interaction of matter and radiation: electromagnetic, strong, weak and gravitational. You will see that these four interactions underlie *all* processes, at *all* scales, *everywhere* in the Universe. Then in Chapter 14, the question is raised of whether the four interactions are truly distinct or whether there might be bigger and better theories, unifying two or more of the four interactions.

Bringing together the information from the earlier chapters, the final part of the book, comprising Chapter 15, presents a history of the Universe, from the earliest times about which it makes sense to speak, through to the present day.

This book contains most of the material that you will need, but there are also a couple of computer-based activities to support your study. The first of these, in Chapter 5, is called *Electrons in Atoms*. It serves to develop your understanding of how quantum rules apply inside atoms and will enable you to see how light is emitted or absorbed by atoms, and the effect of these processes on the atoms themselves. The other computer-based activity, embedded within Chapter 8, is entitled *The Virtual Telescope*. Using your computer you will take measurements of galaxies from which you will be able to measure the expansion rate of the Universe and so calculate its age.

Some of the ideas discussed in this book may challenge the view of the world that you currently hold, and throughout history such challenges have been one of the hallmarks of scientific progress. Later in this book, we'll reflect on these 'paradigm shifts' that have characterised the history of science, many of which you have already encountered earlier in the course. Apart from the intellectual excitement of topics in cosmology and particle physics, they also serve to illustrate the way in which scientists continually strive to push back the boundaries of knowledge, extrapolating from what can be measured in the laboratory to realms that are impossible to study directly. But that is not to say these subjects have no relevance to our everyday lives – for instance, the technological age that we now live in could not have come about were it not for the underpinning science of quantum physics. However, you will also encounter some rather bizarre ideas in the following pages. These will include particles that appear out of nothing, gravitational waves that permeate the entire Universe and state of the art theories (for example, that we live in an eleven-dimensional space-time!). As you will realise, there are some difficult concepts in the material ahead. But don't worry – you are now nearing the end of the course, and whilst the ideas ahead may be challenging, we hope you're going to find them stimulating and fascinating too. Prepare, therefore, for some mental exercise as you embark on a journey to the frontiers of physics and an exploration of the processes that govern the Universe.

If you have not already done so, at this point you should refer to the Study Calendar in order to plan your study time for the rest of the book.

Activity 1.1 Keeping track of symbols and equations

We expect this activity will take you approximately 15 minutes.

As you have worked through this course, various activities have encouraged you to keep track of the various symbols and equations that have been introduced by compiling your own glossary of symbols and their meanings. We suggest you continue to add to your glossary as you work through this book. As well as algebraic symbols for various quantities and equations describing relationships between different quantities, you might also like to include values for the various physical constants that you will meet (such as the speed of light, the Planck constant, the Hubble constant, etc.). We suggest you revisit this activity at the end of each chapter.

There are no comments on this activity.

1.3 Summary of Chapter 1

The phrase 'quarks to quasars' encompasses 45 orders of magnitude in length scale and spans the entire human comprehension of the Universe. The smallest length scales are the realm of particle physics, whilst the largest length scales are the realm of cosmology.

The fields of cosmology and particle physics are linked, as it is only in particle accelerators that scientists can recreate the extreme conditions that existed throughout the Universe during the first moments of its existence.

This book will explore the answers to questions concerning the behaviour of the Universe on small and large scales, the rules which the Universe follows, and how the Universe changes with time throughout its history.

PART 1 – HOW DOES THE UNIVERSE BEHAVE ON SMALL SCALES?

> One sees great things from the valley; only small things from the peak.
>
> G. K. Chesterton, British writer (1874–1936)

As noted in Chapter 1, this part of the book is where the ideas of quantum physics are introduced. Over the past century, scientists have made several great voyages of discovery, but perhaps the most spectacular of them was the one in which they travelled the shortest distance. By penetrating only a tenth of a millionth of a millimetre into atoms, scientists have uncovered a new quantum world with a host of features that no one had predicted. This first part of the book therefore addresses the first part of the question posed in Chapter 1 – how does the Universe behave on small scales?

Chapter 2
Reminders from earlier in the course...

> Right now I'm having amnesia and déjà vu at the same time.
>
> Steven Wright, American comedian (1955–)

In Book 3 you learned about energy and electromagnetic radiation and in Book 4 you saw the basic picture of how atoms are built. In this chapter, you will revisit some of the key facts that you learned to help your understanding of what follows. The discussion of these concepts is extended a little, and some questions are posed that are raised by what you have already learned, each of which will be answered in the following chapters.

2.1 The world of atoms

As you learned in Book 4, everything around us is composed of atoms, and there are known to be around 90 different types of atom that occur naturally in the world. A material made of a single type of atom is known as an element. The most abundant elements in the Universe as a whole are those composed of the two simplest atoms: hydrogen and helium. Later in this book, you will see just why this is the case. Here on Earth, there are also significant amounts of other elements, in particular carbon, nitrogen, oxygen, sodium, magnesium, aluminium, silicon, sulfur, calcium and iron. Later on, you will find out where these elements came from.

Whatever the type of atom, each one has certain features in common. Each contains a central nucleus, which carries a positive electric charge as well as most of the atom's mass. The nucleus is surrounded by one or more negatively charged electrons (symbol e^-) each of which has a much lower mass than the nucleus. The nucleus of an atom is what determines the type of element. The very simplest atoms of all, those of the element hydrogen, have a nucleus consisting of just a single proton (symbol p). The next simplest atom, helium, has two protons in its nucleus; lithium has three protons; beryllium has four; boron has five; carbon has six; and so on. The *number* of protons in the nucleus of an atom is known as its atomic number (Z). As you know, elements have very different chemical properties. Later in this book, you will see why this is so.

The electric charge (Book 3, Section 7.1) of a proton is represented by the algebraic quantity '$+e$' with a numerical value of 1.6×10^{-19} coulombs (symbol C) (to two significant figures). The electric charge of an electron is exactly the same as that of a proton, but negative instead of positive, so is written as '$-e$', which has a value of -1.6×10^{-19} C.

■ What is the atomic number of carbon? What is the electric charge of a carbon nucleus?

□ The nucleus of a carbon atom contains six protons, so the atomic number of carbon is 6 and the charge of the nucleus is $+6e$, which is equivalent to $6 \times (1.6 \times 10^{-19}\ \text{C}) = 9.6 \times 10^{-19}$ C. To the nearest order of magnitude, this is therefore 10^{-18} C.

The other constituents of atomic nuclei are neutrons (symbol n) which have a similar mass to protons, but have zero electric charge. Normal hydrogen atoms have no neutrons in their nuclei, although there is a form of hydrogen – known as deuterium – that does. The nucleus of a deuterium atom consists of a proton *and* a neutron. It is still the element hydrogen (since it contains one proton) but it is a 'heavy' form of hydrogen, thanks to the extra neutron. Deuterium is said to be an isotope of hydrogen. Similarly, normal helium atoms contain two neutrons in their nucleus, along with the two protons, but a 'light' isotope of helium, known as helium-3, contains only one neutron instead. The combined number of protons and neutrons in the nucleus of an atom is the mass number (symbol A) of the atom; essentially this is the relative atomic mass of a particular isotope. Isotopes therefore denote forms of the same element with different mass numbers. The nucleus of a particular isotope is referred to as a nuclide.

- ■ What are the mass numbers of (a) normal hydrogen (b) 'heavy' hydrogen (i.e. deuterium) (c) normal helium and (d) 'light' helium (i.e. helium-3)?

- □ (a) The nucleus of normal hydrogen contains one proton and no neutrons, so the mass number is 1. (b) The nucleus of 'heavy' hydrogen contains one proton and one neutron, so the mass number is 2. (c) The nucleus of normal helium contains two protons and two neutrons, so the mass number is 4. (d) The nucleus of 'light' helium contains two protons and one neutron, so the mass number is 3.

As a short-hand, isotopes of each atomic element may be represented by a symbol. Letters are used to indicate the name of the element itself, and two numbers are used to indicate the atomic number (lower) and mass number (upper). So a normal hydrogen atom is represented as ${}^{1}_{1}\mathrm{H}$, and an atom of the heavier isotope, deuterium, by ${}^{2}_{1}\mathrm{H}$. Isotopes of some other light atoms are indicated in Figure 2.1.

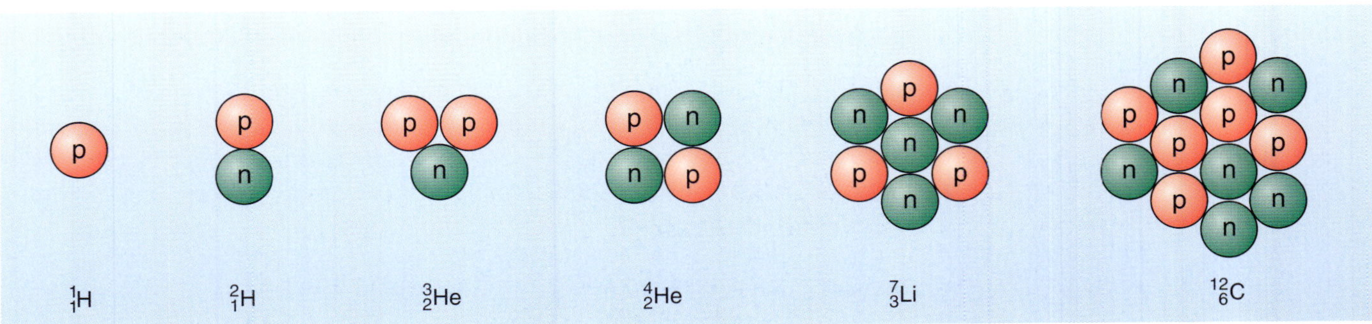

Figure 2.1 Schematic diagrams of the nuclei of some isotopes. Protons, p, are coloured red, and neutrons, n, are coloured green, in a convention that will be followed throughout this book.

Sometimes, protons and neutrons are collectively referred to as **nucleons** since both types of particle are found inside the nucleus of an atom. Similarly, electrons, protons and neutrons are often collectively referred to as subatomic particles, for obvious reasons. This nomenclature is summarised in Table 2.1.

Normal atoms are electrically neutral, so the positive electric charge of the nucleus is exactly balanced by the negative electric charge of the electrons

Table 2.1 The constituents of atoms.

Subatomic particles		
electron, e^- 	nucleons mass number = total number of nucleons in the nucleus any specific type of nucleus is called a nuclide	
In a neutral atom, number of electrons = number of protons	proton, p  atomic number = total number of protons	neutron, n  isotopes of the same element have different numbers of neutrons
electric charge = $-e$	electric charge = $+e$	electric charge = 0

surrounding it. Since each electron carries an electric charge of $-e$ and each proton carries an electric charge of $+e$, the number of electrons in a neutral atom is *exactly* the same as the number of protons in its nucleus.

■ What is the difference between an atom of lithium-7 and an atom of beryllium-7? (*Hint*: if you need a reminder of the atomic numbers for these elements, refer back to the Periodic Table in Book 4 Figure 5.3.)

□ Both atoms have the same mass number, namely 7. However, the nucleus of the lithium atom has 3 protons and 4 neutrons, whilst the nucleus of the beryllium atom has 4 protons and 3 neutrons. Furthermore, the lithium atom contains 3 electrons whilst the beryllium atom contains 4 electrons.

■ The element iron has an atomic number 26, and its most common isotope is known as iron-56. (a) How many protons and how many neutrons are there in a single nucleus of iron-56? (b) How many electrons are there in an electrically neutral atom of iron-56?

□ (a) Since the atomic number is 26, the nucleus contains 26 protons. Since the mass number is 56, the total number of protons and neutrons in the nucleus is 56, and so the nucleus contains $56 - 26 = 30$ neutrons. (b) An electrically neutral atom contains the same number of electrons as protons, so the atom contains 26 electrons.

As you saw in Book 4 (Section 5.1), chemists find it convenient to think of the electrons in an atom as being arranged in a series of concentric shells and subshells surrounding the nucleus. Whilst this description helps explain why certain atoms interact with others in particular ways, it is a model that only partially explains the behaviour of atoms. You might be asking yourself, *why* do electrons arrange themselves in this way? In the following chapters you will see how the shell picture emerges.

Finally, here is a reminder of the size of atoms and nuclei. Whereas a typical atomic nucleus has a size of around 10^{-14} m, the size of the atom itself is determined by the size of the region occupied by the electrons that surround the nucleus. The overall size of an atom is about 10^{-10} m across.

2.2 The world of energy

If there is one equation in physics that people have heard of, it is Einstein's famous statement:

$$E = mc^2 \tag{2.1}$$

As you saw in Book 3 (Section 8.2), what Einstein tells us via this equation is that energy (E) and mass (m) are interchangeable. Energy may be converted into mass and mass may be converted into energy. The conversion factor linking the two is the speed of light squared (c^2, where $c = 3.0 \times 10^8$ m s^{-1}).

Mass is a physical property that quantifies the amount of matter in a body. If, for simplicity, you consider an object like a diamond that is composed entirely of carbon atoms, then a more-massive diamond will contain more atoms of carbon than a less massive diamond does.

The conservation of energy is a fundamental principle of the Universe, and one definition of energy is simply that it is the 'stuff' that is conserved during any physical process. Alternatively, energy can be said to be a physical property possessed by an object that quantifies its capacity to make changes to other objects. There are a variety of possible changes, and these include changes in speed of motion, changes in temperature and changes in position with respect to other massive or electrically charged objects. Perhaps the most familiar form of energy is kinetic energy (Book 3 Chapter 3), or energy of movement. Some other types fall into the category of potential energy, namely energy that is stored and which depends on the position of an object. Gravitational energy (Book 3 Section 5.3) and electrical energy (Book 3 Section 7.4) are two forms of potential energy that you have met before. Electromagnetic radiation, such as light, also carries energy (Book 3 Section 10.2).

Although the SI unit of energy is the joule (symbol J), when dealing with atoms and subatomic particles, the most convenient unit to use for energy is the electronvolt (symbol eV). As you saw in Book 3 (Section 7.4), the conversion factor between the two units is that 1 eV is equal to 1.6×10^{-19} J (to 2 significant figures), which happens to be the right sort of size to describe energies at the atomic and subatomic scales.

Now, because mass and energy are interchangeable, it is also convenient to refer to the masses of subatomic particles in terms of their energy equivalence. For instance, the mass of an electron is 9.1×10^{-31} kg (to two significant figures). So the energy equivalent of this mass is given by Equation 2.1 as:

$$E = (9.1 \times 10^{-31}\ \text{kg}) \times (3.0 \times 10^8\ \text{m s}^{-1})^2 = 8.2 \times 10^{-14}\ \text{J}.$$

Converting this into electronvolts:

$$E = \frac{8.2 \times 10^{-14}\ \text{J}}{1.6 \times 10^{-19}\ \text{J eV}^{-1}} = 5.1 \times 10^5\ \text{eV to 2 significant figures}$$

So the energy equivalent of the mass of an electron is around 510 keV (510 kiloelectronvolts or 510 thousand electronvolts). The term **mass energy** is often used to refer to this energy equivalent of the mass of a particle.

■ Given that the mass of either a proton or neutron is about 1.7×10^{-27} kg, follow a similar procedure to that above to calculate their approximate mass energy in eV.

□ The mass energy is given by $E = (1.7 \times 10^{-27}\ \text{kg}) \times (3.0 \times 10^{8}\ \text{m s}^{-1})^2 = 1.5 \times 10^{-10}$ J. Converting this into electronvolts:

$$E = \frac{1.5 \times 10^{-10}\ \text{J}}{1.6 \times 10^{-19}\ \text{J eV}^{-1}} = 9.4 \times 10^{8}\ \text{eV to two significant figures}$$

So the mass energy of either a proton or neutron is around 940 MeV (940 megaelectronvolts or 940 million electronvolts).

Equation 2.1 may also be rearranged as $m = E/c^2$, so masses may be expressed in units of 'energy/c^2'. The approximate masses of an electron and a proton or neutron can therefore be written as 510 keV/c^2 and 940 MeV/c^2 respectively. A summary of the masses and mass energies of the electron, proton and neutron is given in Table 2.2 for reference.

Table 2.2 Approximate masses and mass energies of the constituents of atoms.

Particle	Approximate mass		Approximate mass energy
proton, p	1.7×10^{-27} kg	940 MeV/c^2	940 MeV
neutron, n	1.7×10^{-27} kg	940 MeV/c^2	940 MeV
electron, e^-	9.1×10^{-31} kg	510 keV/c^2	510 keV

■ Given the information in Table 2.2, how much more massive than an electron is a proton?

□ The mass of a proton is 940 MeV/c^2 or 9.40×10^8 eV/c^2, whilst the mass of an electron is 510 keV/c^2 or 5.10×10^5 eV/c^2. The ratio of the masses is:

$$\frac{9.40 \times 10^{8}\ \text{MeV}/c^2}{5.10 \times 10^{5}\ \text{MeV}/c^2} = 1840\ \ \text{(to three significant figures)}$$

so a proton is almost 2000 times more massive than an electron.

Taking the idea of mass energy equivalence a little further, two particular ways in which mass and energy can be converted from one to the other are via the processes called pair production and matter–antimatter annihilation. Although you may have thought **antimatter** to be the stuff of science fiction, it is a very real feature of the Universe. Antimatter particles have the same mass as their matter counterparts but their other attributes, such as electric charge, have the opposite sign. All matter particles have corresponding antimatter counterparts. Our Universe today seems to consist almost exclusively of matter particles rather than antimatter. However, as you will see later, the early Universe was not such a one-sided place.

The antimatter counterpart of the electron, known as the **positron** (or **antielectron**), was discovered in 1932. More recently, in 1996, atoms of antihydrogen were created, consisting of **antiprotons** bound to positrons. Nowadays antimatter particles can be created routinely in high-energy particle accelerators, but antimatter is difficult stuff to control. If matter and antimatter come into contact with each other they will mutually annihilate producing a large amount of energy, which appears in the form of electromagnetic radiation. The process of **matter–antimatter annihilation** may therefore be expressed as:

$$\text{matter} + \text{antimatter} \longrightarrow \text{electromagnetic radiation} \qquad (2.2a)$$

The process of **pair creation** is exactly the reverse:

$$\text{electromagnetic radiation} \longrightarrow \text{matter} + \text{antimatter} \qquad (2.2b)$$

As you saw in Book 3 Chapter 2, one of the most important rules governing how the Universe works is the principle of conservation of energy.

■ State the principle of conservation of energy.

□ Energy cannot be created or destroyed, but merely *changed* from one form to another.

So, for instance, any amount of kinetic energy, electrical energy, gravitational energy, mass energy, or energy of electromagnetic radiation may be converted into exactly the same amount of any other type of energy. Notice that this principle *includes* mass energy as a form of energy, just like any other.

In the annihilation and pair creation reactions, you have to be careful though, because there are *two* types of energy to consider when talking about the matter and antimatter. First, there is the mass energy of the various matter and antimatter particles and, in addition to this, there is the kinetic energy possessed by the particles. So, in annihilation reactions, the energy of the electromagnetic radiation that is produced is equal to the combined mass energy of the matter and antimatter particles *plus* their combined kinetic energy. Similarly, in pair creation reactions, the energy of the electromagnetic radiation can appear as the combined mass energy of the matter and antimatter particles, and any energy left over is imparted to the particles as kinetic energy.

■ An electron and a positron *at rest* annihilate each other. How much energy is carried by the electromagnetic radiation that is created?

□ The electron and positron each have a mass energy of about 510 keV. Since they are at rest, their kinetic energy is zero. The total energy available is therefore 2×510 keV, so the energy carried by the electromagnetic radiation is just over 1 MeV.

■ If a proton and antiproton are created from electromagnetic radiation, what is the *minimum* total energy that the radiation must have?

□ A proton and antiproton each have a mass energy of about 940 MeV, so the radiation must have an energy of at least 1880 MeV (i.e. almost 2 GeV). If the electromagnetic radiation had more energy than this, then the excess would appear as kinetic energy of the proton and antiproton.

2.3 The world of light

Our eyes are sensitive to light that comprises the familiar rainbow of colours, but as you saw in Book 3 (Section 8.6), these colours are merely a tiny part of the vast electromagnetic spectrum. In this section, you are reminded of the key features of light that you learned about earlier in the course.

When a beam of white light, such as that from the Sun, is passed through a glass prism, it is broken up, or dispersed, into a band of colours that is called the visible spectrum (Figure 2.2). The spectrum of sunlight contains visible light of all colours and therefore forms a continuous spectrum. Some other objects also emit continuous spectra of light. For instance, the heated metal of a tungsten filament in a light bulb, or the hot-plate on an electric cooker, each emit a continuous spectrum.

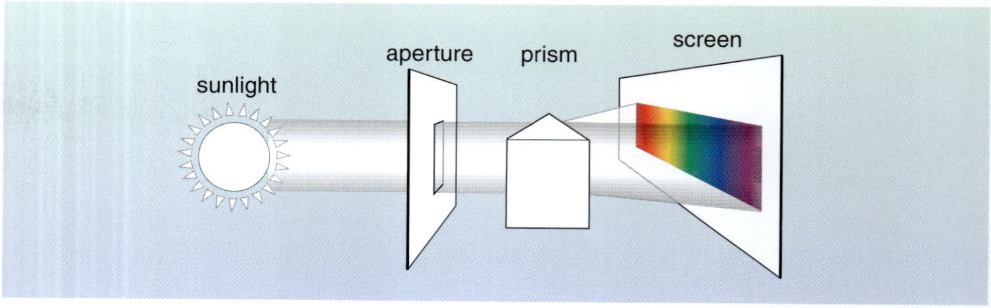

Figure 2.2 Using a glass prism to view the spectrum of a beam of white light.

The reason that some things appear to be coloured is that atoms and molecules absorb and emit different colours of light.

■ Why do you suppose that the petal of a rose appears red when sunlight shines on it?

□ The petal absorbs all the visible colours *except* the red light. Some of the red light passes through the petal, and some is reflected from its surface.

■ Why do you think a piece of paper appears white when viewed in sunlight?

□ Sunlight has a continuous spectrum composed of all colours. All of these colours are reflected equally well by the paper.

■ What do you suppose is the reason that water and glass appear essentially colourless when sunlight shines through them?

□ Water and glass do not significantly absorb any colours, so all colours of light are transmitted through.

It turns out that every type of atom shows a preference for absorbing or emitting light of certain *specific* colours. So each atom is associated with a characteristic pattern of colours – a sort of technicolour fingerprint. Just as every human being has their own set of characteristic fingerprints, so every type of atom has its own pattern of colours of light that it can absorb or emit.

The yellow glow produced by 'sodium' street lights when an electric current passes through vaporised sodium is the fingerprint of that element; similarly, the bright orange–red light of a neon sign is the fingerprint of the element neon. Such associations are even more apparent when a prism is used, in an arrangement similar to that of Figure 2.2, to disperse the light from a source of known chemical composition. In this case, spectra consisting of emission lines are seen; Figure 2.3b shows the emission spectrum of sodium. Exactly the same association may be seen by observing how white light is *absorbed* by sodium atoms. This is done by passing a beam of white light through some sodium vapour and examining the spectrum of the emerging beam to see if any colours are missing, as shown in Figure 2.3a. Figure 2.3c shows the resulting absorption spectrum of sodium; as you can see it exhibits a dark absorption line, marking the absence (due to absorption by the sodium vapour) of exactly the same shade of yellow that was seen in the emission spectrum.

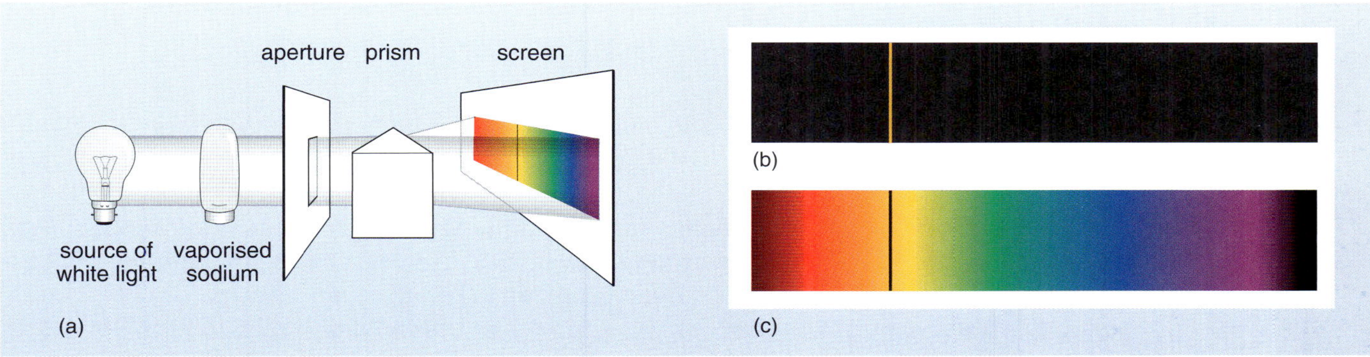

Figure 2.3 (a) An arrangement that may be used to display the absorption spectrum of a particular substance, in this case sodium vapour. (b) The emission spectrum of sodium in the visible part of the electromagnetic spectrum. (c) The absorption spectrum of sodium in the visible part of the electromagnetic spectrum.

As you saw in the experiment you carried out in Book 3 Activity 11.1 where you measured spectra for yourself, sodium and neon atoms are by no means unique in having a characteristic pattern of spectral lines. In fact, every kind of atom has an associated, characteristic 'spectral fingerprint'. Figure 2.4 shows the visible emission spectra of a few chemical elements. In each case the light being examined originates from a lamp that contains a huge number of atoms of the relevant type, and the observed spectrum is characteristic of every atom of that type. For this reason, the kind of emission spectra shown in Figure 2.4 are often called atomic spectra.

One way to quantify the colours in the spectral fingerprint of each element is to use the wave description of light. Whilst it is fairly easy to visualise water waves, or even sound waves, it is not immediately apparent what is 'waving' in the case of a light wave. This will be discussed further in Chapter 10, but for now simply recall (Book 3 Chapter 9) that the distance between one crest of a wave and the next is known as the wavelength of the wave, usually represented by the Greek letter lambda, λ. The frequency of a wave, usually represented by the letter f, is the number of cycles of a wave that pass a fixed point per second.

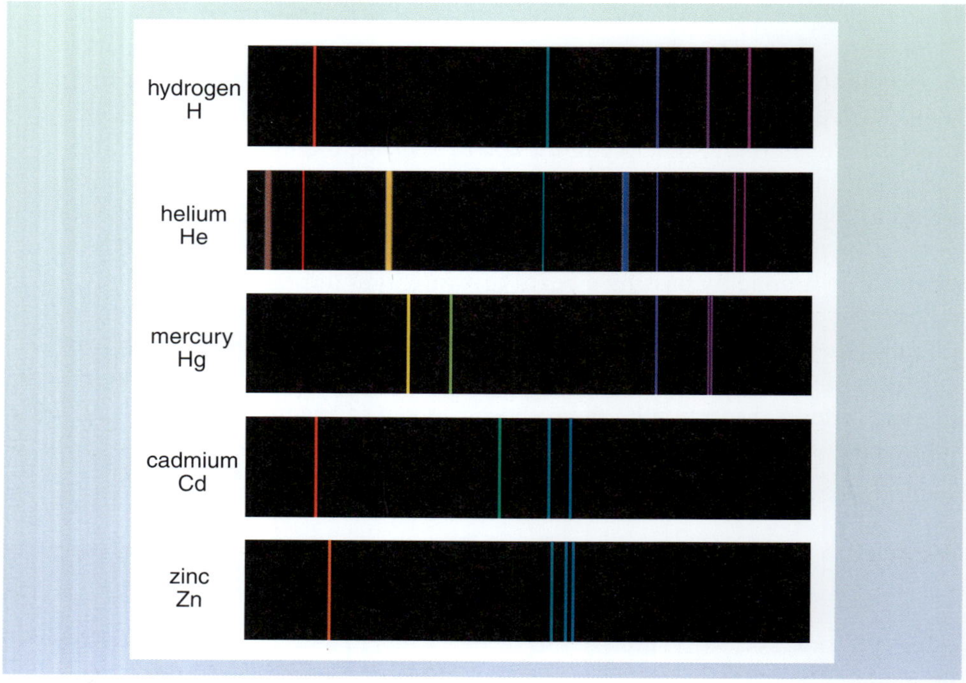

Figure 2.4 The emission spectra that characterise a number of elements (hydrogen, helium, mercury, cadmium and zinc). Note that only the visible spectra are shown here: all these atoms also emit radiation that is not visible.

Electromagnetic radiation travels at the speed of light, c, which has a value of 3.0×10^8 m s^{-1} (to two significant figures) in a vacuum, and the speed is related to the frequency and wavelength of the radiation by the equation:

$$c = f\lambda \tag{2.3}$$

The key thing to remember is that each colour of light corresponds to electromagnetic waves of a different wavelength. Visible light spans the range of wavelengths from about 400 billionths of a metre (400 nm or 400×10^{-9} m) to 700 billionths of a metre (700 nm or 700×10^{-9} m). Violet light has the shortest wavelength and red light the longest wavelength (Figure 2.5). However, at even longer wavelengths are found first infrared radiation, then microwaves and then radio waves, whilst at wavelengths shorter than the visible are found ultraviolet radiation, then X-rays and then gamma rays. The full electromagnetic spectrum is shown in Figure 2.6.

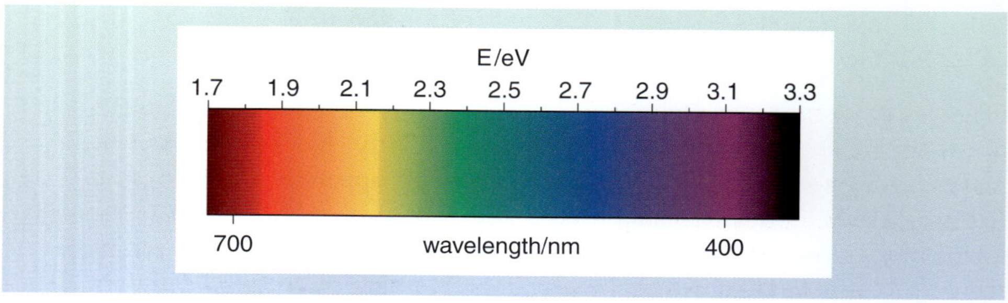

Figure 2.5 Different colours of light correspond to radiation of different wavelengths and energies.

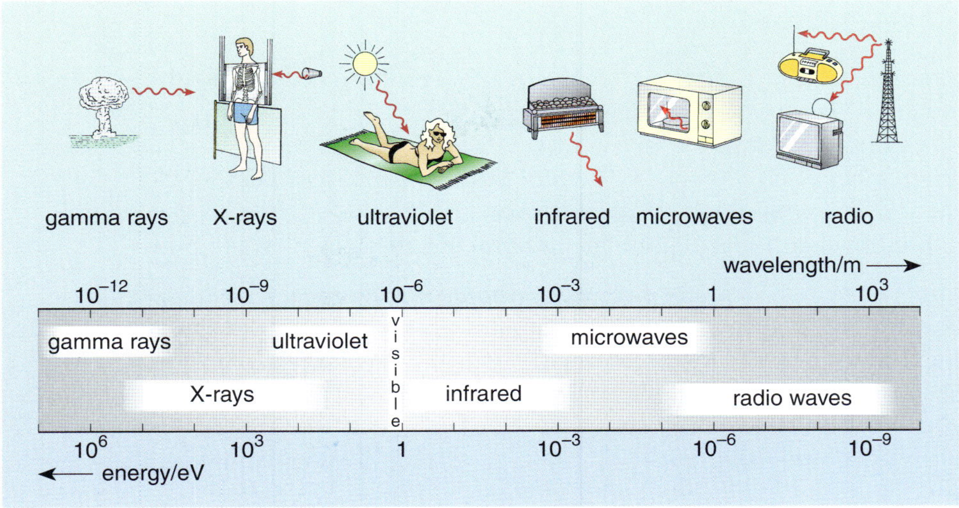

Figure 2.6 The complete electromagnetic spectrum.

Although electromagnetic radiation travels from place to place like a wave, it is emitted or absorbed by matter as if it is composed of particles. Monochromatic light, which has a single colour, consists of identical particles, called photons, each of which has exactly the same energy.

The unit of energy used to quantify a photon's energy is the electronvolt, and conveniently photons of red light each have an energy of just less than 2 eV, whilst photons of violet light each have an energy of just over 3 eV. At lower energies than red light are the infrared, microwave and radio wave parts of the electromagnetic spectrum, and at higher energies are the ultraviolet, X-ray and gamma-ray regions. As you can see on Figure 2.6, X-ray or gamma-ray photons have energies of many keV (thousands of eV) or many MeV (millions of eV).

There is no conflict between the two approaches to describing light; it is just that one picture (waves) is useful for describing the way electromagnetic radiation propagates and another picture (particles) is useful for describing the way it interacts with matter. This double description is referred to as wave–particle duality.

The energy of a photon, E_{ph}, is related to the frequency, f, used to characterise the electromagnetic wave by the equation (Book 3 Section 10.2):

$$E_{\text{ph}} = hf \tag{2.4}$$

The symbol h represents the Planck constant which has a value of 4.1×10^{-15} eV s (to two significant figures). In this sense, the Planck constant can be thought of as a simple conversion factor between photon energy and electromagnetic wave frequency. More fundamentally, the Planck constant lies at the heart of understanding the behaviour of matter on its smallest scales. Its introduction in the late 19th and early 20th centuries was (literally) a symbol of the revolution in scientific thinking that was necessary for scientists to tackle the problem of understanding the true nature of matter and how it interacts with electromagnetic radiation. The revolution was that of quantum physics, which you will be learning a good deal about in the next few chapters.

2.4 Summary of Chapter 2

Atoms consist of a positively charged nucleus surrounded by negatively charged electrons. Each type of atom is known as an element and is distinguished by its atomic number (Z), which is simply the number of protons in its nucleus, or the number of electrons in the neutral atom. The mass number (A) of an atom is the total number of protons and neutrons in its nucleus. Different isotopes of a particular element have different numbers of neutrons.

Energy can take a variety of forms including kinetic energy, electrical energy and gravitational energy. In all physical processes, the total energy is conserved. Furthermore, mass (m) and energy (E) may be converted from one to the other according to Einstein's most famous equation, $E = mc^2$, where c is the speed of light. Mass energy may therefore be considered as simply another form of energy. A convenient unit of energy to use in atomic processes is the electronvolt (eV), and a convenient unit of mass is eV/c^2. The mass energy of an electron or positron is about 510 keV, whilst the mass energy of a proton or neutron is about 940 MeV, which is almost 2000 times larger. The processes of matter–antimatter annihilation and pair production allow mass to be converted into energy and energy to be converted into mass, respectively.

Light and other electromagnetic radiation travels from place to place as if it were a wave, characterised by its wavelength or frequency. It interacts with matter as if it were composed of particles called photons, each characterised by a specific energy. The key equations relating the wavelength (λ) and frequency (f) of electromagnetic radiation to photon energy (E_{ph}) are: $c = f\lambda$ and $E_{\text{ph}} = hf$, where c is the speed of light ($3.0 \times 10^8\ \text{m s}^{-1}$) and h is the Planck constant ($4.1 \times 10^{-15}\ \text{eV s}$).

The spectrum of electromagnetic radiation emerging from an object may take the form of a continuous spectrum, or it may include emission or absorption lines of specific wavelengths (or energies). The characteristic pattern of spectral lines arising from a particular element is a spectral fingerprint, which is unique to that particular type of atom.

Chapter 3
Quantised energy

> Common sense is the deposit of prejudice laid down in the mind before the age of eighteen.
>
> Albert Einstein, German-Swiss-American physicist (1879–1955)

One of the most revealing ways of investigating atoms is to look at how they interact with light. By examining the light emitted and absorbed by atoms, scientists have been able to draw up an astonishingly detailed picture of how the electrons in atoms are arranged. In this chapter you will see that, since atoms are characterised by the light they emit and absorb, it is possible to work out, from the energies of this light, the possible values of energy that the atom can gain or lose.

Figure 3.1 Max Planck (1858–1947) began the quantum revolution. At the time, he was one of Germany's leading physicists and, as an exceptionally conservative scientist, was an unlikely revolutionary. However, he quickly appreciated the importance of his quantum idea: his son Erwin recalled that Planck said to him during a walk in a park in late 1900, 'Today I have made a discovery which is as important as that of Newton'. He was not exaggerating. On that occasion Planck introduced the first of the many concepts that now make up quantum physics.

3.1 The quantum world

If you had told most physicists at the end of the 19th century that their branch of science was going to undergo a complete upheaval, they would probably not have believed you. At that time, there was a widespread, though not universal, complacency in the scientific community. As early as 1875, a young student called Max Planck began his studies at the University of Munich, only to be encouraged by the professor of physics not to study science as there was nothing new to be discovered! Planck (Figure 3.1) wisely ignored his professor's advice and went on to make what is now regarded as one of the most important discoveries in the entire history of science, a discovery that has been fundamental to our modern understanding of atoms.

Towards the end of the 19th century, many scientists were trying to understand the properties of the radiation emitted by hot objects. *All* objects emit electromagnetic radiation, and the amount of radiation that they produce at different wavelengths depends on the temperature of the object. For instance, the electromagnetic radiation emitted by the Earth is mainly in the form of infrared radiation, whereas that from a hotter body, such as the Sun, is mainly in the visible part of the spectrum. Some people thought this phenomenon was a minor diversion and that it would be understood using old ideas, but in it laid the seeds of a scientific revolution.

■ A scientific revolution is a period marked by some major discoveries that lead to a complete change in the way an area of science is approached. These are sometimes called **paradigm shifts**. Can you think of any other scientific revolutions that you have met so far in this course?

□ Other examples are: Newton's laws which led to a new way of understanding motion both on Earth and in the Solar System (Book 2); plate tectonics which led to a revolution in the understanding of the Earth (Book 2); Darwin's theory of evolution by natural selection (Book 5) and Mendeleev's construction of the Periodic Table of the elements which revolutionised understanding of chemical processes (Book 4).

In 1900, Planck made the astonishing suggestion that when objects emit or absorb radiation they can do so only in multiples of a certain minimum amount of energy called a **quantum** (plural quanta, from the Latin word *quantus* meaning 'how much'). So, for example, the gas inside an energy-saving light bulb emits energy in the form of light, but it does not emit energy in just any amounts, rather it emits it in certain, particular amounts: quanta of light. Likewise, when light is absorbed by, say, a piece of metal, the energy of the light is absorbed only in quanta. The details of Planck's idea need not concern you; what is important for the present purposes is the historical significance of his idea. Planck was the first to introduce the idea of the quantum into science, but within 25 years, other scientists had used it to form the basis of quantum physics, which in turn is the basis of our current understanding of how atoms behave.

Perhaps the culmination of these early developments in quantum physics came at the 5th Solvay International Conference, held in Belgium in 1927. The topic of the meeting was *Electrons and Photons*, and amongst the 32 delegates (29 of whom appear in the photograph in Figure 3.2) were 18 scientists who were,

Figure 3.2 The delegates of the 1927 Solvay International Conference, held in Brussels. In the middle of the back row stands Erwin Schrödinger (Nobel Prize 1933); two places to the right of him (from your perspective) are Wolfgang Pauli (Nobel Prize 1945) and Werner Heisenberg (Nobel Prize 1932). Seated in the centre of the middle row is Paul Dirac (Nobel Prize 1933); two places to the right of him is Louis de Broglie (Nobel Prize 1929), whilst at the right-hand end of the row is Niels Bohr (Nobel Prize 1922). Finally, in the front row, Max Planck (Nobel Prize 1918) sits to the left of Marie Curie (Nobel Prizes 1903 and 1911), and two places to the right of her is Albert Einstein (Nobel Prize 1921). The scientists mentioned here by name are each referred to later in the book, and you will revisit this photograph in Activity 7.1.

or became, Nobel Prize winners (Marie Curie actually won two – one each for physics and chemistry). Many of these scientists feature in the next few chapters of this book.

This scene-setting for your exploration of quantum physics concludes with a consideration of the vital importance of this subject for understanding and describing nature. Ask an astronomer why stars, such as the Sun, are able to shine so brightly, and you will learn that the answer lies ultimately in the realm of quantum physics. Ask a chemist why some atoms stick together to form molecules but others do not, or ask a biologist why DNA molecules fold up in the particular way that they do, and you will again be told that the answer ultimately lies in quantum physics. Ask almost anyone how the whole Universe came into existence and they will probably say that they haven't a clue; but if you speak to one of the small band of scientists who are actively engaged on this problem they will tell you that their best bet is that it involves quantum physics in some fundamental way.

Quantum physics explains how the fundamental particles of matter and radiation interact with one another. It explains how and why electrons are arranged the way they are in atoms, and this in turn provides an explanation for the various properties of the different elements. Ultimately, therefore, all of the rich diversity of chemical reactions relies on the rules of quantum physics. Taken further, the chemical reactions that underlie biological systems, such as the transcription of DNA and the process of glucose oxidation, also rely on the underpinning rules of quantum physics. Some scientists even suggest that the very essence of human consciousness may lie in quantum processes occurring within our brains.

At the end of Chapter 3, you will return to the theme of the relevance of quantum physics to everyday life and look at how quantum physics has enabled the technology that many of us now take for granted.

3.2 The spectral fingerprint of hydrogen

What can photons reveal about the inner workings of an atom? Well, each spectral line involves the emission or absorption of photons of a particular energy. So it follows that in order to explain why each type of atom is associated with a particular set of spectral lines, what is really needed is an explanation of why each type of atom only emits or absorbs photons that have certain fixed amounts of energy. These photon energies are just as characteristic of a given type of atom as the pattern of spectral lines to which they correspond. This is a key point:

> Each type of atom can be characterised by the energies of the photons that it can absorb or emit.

The message, then, is simple: every atom has a spectral fingerprint that can be specified in terms of either the colours of the radiation it emits or absorbs or, more precisely, the energies of the corresponding photons.

You will now apply these ideas to the simplest atom of all, the hydrogen atom.

■ The hydrogen emission spectrum shown in Figure 2.4 contains five visible lines that characterise the hydrogen atom. Using the energy scale from Figure 2.5, estimate the five corresponding photon energies (in electronvolts) that also characterise the hydrogen atom.

□ The values for the five spectral lines in the visible part of the hydrogen spectrum are approximately 1.9 eV (red), 2.6 eV (blue–green), 2.9 eV (deep blue), 3.0 eV (violet) and 3.1 eV (also violet).

What do the five spectral lines in the visible part of the spectrum imply for the hydrogen atom? Well, you have just seen that the spectral fingerprint of hydrogen atoms can be specified by five particular energies, which are 1.89 eV, 2.55 eV, 2.86 eV, 3.02 eV and 3.12 eV when expressed to three significant figures. This set of energy values represents the experimental fact that hydrogen atoms emit light only with these five values of energy (likewise, they can absorb light only with these five energy values too). So, for example, a hydrogen atom can emit a photon with an energy of exactly 1.89 eV, but can never emit a photon with an energy of, say, 1.91 eV. The hydrogen atom, then, is very particular about the light it emits and absorbs. The same thing is true for all other atoms – every type of atom can emit or absorb photons with only particular values of energy.

3.3 The energy of an atom

You know that energy is always conserved. This implies that if an atom *absorbs* a photon that has a given energy, then the energy of that atom must *increase* by the same amount of energy. Similarly, if an atom *emits* a photon of a particular energy, then the energy of that atom must *decrease* by the same amount of energy. So, the fact that a given type of atom can absorb and emit photons of certain, precisely defined, energies must mean that the atoms themselves are only able to increase and decrease their own energy by exactly those amounts.

■ (a) A helium atom absorbs a photon of energy 2.11 eV. What happens to the energy of the helium atom as a result? (b) A mercury atom emits a photon of energy 2.27 eV. What happens to the energy of the mercury atom as a result?

□ (a) If a helium atom absorbs a photon of energy 2.11 eV, then the energy of the atom must *increase* by exactly this amount. (b) If a mercury atom emits a photon of energy 2.27 eV then the energy of the atom must *decrease* by exactly this amount.

So what is meant by 'the energy of an atom'? Firstly, you should not confuse this with the kinetic energy of the atom as a whole. The atom may well be moving, and so possess a certain amount of kinetic energy, but it is not changes in this energy that are responsible for atomic spectra. Rather your concern here is with what may be thought of as the *internal energy* of the atom. Recall from Book 3 (Section 6.1) that the internal energy of a liquid, say, has two components: the *kinetic energy* of the random motion of the constituent molecules and their *potential energy* that results from the molecular forces of attraction between the molecules. A similar situation applies in the case of individual atoms. The simplest case, a hydrogen atom, consists of a single electron bound to a single proton. The electron will have a certain amount of kinetic energy (because it is moving) and there will be a certain amount of electrical energy due to the electrical force of attraction between the negatively charged electron and the positively charged proton. The *total* energy of the atom is the sum of these two contributions.

Now, think about what will happen if the electron in a hydrogen atom is moved further away from the proton. In Book 3 (Section 5.3), you saw that as an object is raised above the surface of the Earth, work is done against the gravitational force of attraction between the object and the Earth. As a consequence of this, the object acquires increased gravitational potential energy. Here, as the electron is moved further from the proton, work is done against the electrical force of attraction between the two. So in this case, there is an increase in *electrical* potential energy. At the same time, when an electron is moved further from the proton it will generally move more slowly, so its *kinetic* energy will decrease. In fact, the electrical energy generally has a larger magnitude than the kinetic energy, so the electrical energy dominates the energy of the atom. In general, therefore, the further away the electron is from the proton in a hydrogen atom, the greater will be the energy of the atom.

So when a hydrogen atom absorbs or emits photons, the energy of the atom will alter and changes will occur to the internal structure of the atom. In particular, the position and velocity of the electron will change. When a photon is *absorbed*, the position and velocity of the electron will change so that the atom has a *higher* energy, and when a photon is *emitted*, the position and velocity of the electron will change so that the atom has a *lower* energy.

As you will see in Chapter 4, the whole question of the position and velocity of an electron in an atom is subject to so-called quantum indeterminacy, so the picture outlined above is only a guide to the true situation. None the less it is a useful one to bear in mind, and similar conclusions apply to more complex atoms.

3.4 Energy levels and transitions

The ability of any given type of atom to absorb or emit only certain characteristic amounts of energy is explained by saying that the atom itself can have only certain values of energy, known as **energy levels**. The energy is said to be *quantised*, since the energy levels correspond to only certain, well-defined, values of energy. An atom with a particular value of energy, corresponding to a certain energy level, can change to another energy level by absorbing or emitting a photon (there are other ways too, for example by colliding with other atoms), and since energy is conserved, the energy of the photon is precisely equal to the energy difference between the two energy levels. When the atom makes such a change it is said to undergo a **transition** to a different energy level. Such transitions are often informally referred to as *quantum jumps*. This is the first of two key features of the quantum world of atoms:

Each type of atom has a different set of quantised energy levels. When atoms emit or absorb photons, they make transitions between these quantised energy levels.

The two basic types of transition – photon absorption and photon emission – are shown in Figure 3.3. This figure introduces the way that energy levels are represented diagrammatically, and you should study it carefully.

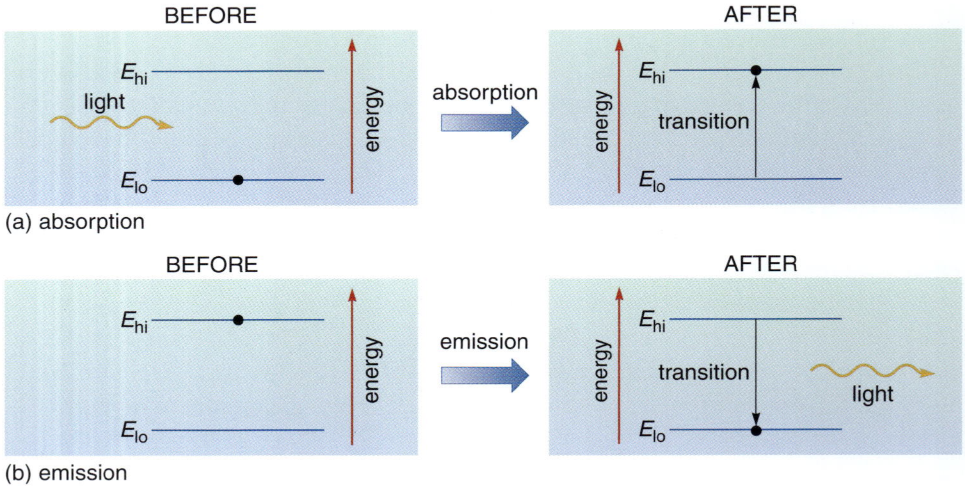

Figure 3.3 The two horizontal lines labelled E_{hi} and E_{lo} in each part of this figure represent two of the possible energies of an atom – two of the possible energy levels. When the atom has a particular value of energy corresponding to a certain energy level, the energy level is marked with a dot. (The widths of the horizontal lines in this diagram are of no significance.) (a) Absorption of a photon. The atom is initially in the lowest energy level E_{lo}; it then absorbs a passing photon and makes a transition to the higher energy level E_{hi}. The photon's energy matches precisely the difference between E_{hi} and E_{lo}. (b) Emission of a photon. Here the atom is initially in the higher level E_{hi}; it emits a photon and makes a transition to the lower level E_{lo}. Again the photon's energy matches precisely the difference between E_{hi} and E_{lo}.

The emission and absorption spectra that you saw earlier arise simply as a result of these transitions between energy levels. When a large number of sodium atoms, for instance, make transitions from a certain high energy level (say, E_{hi}) to a certain *lower* energy level (E_{lo}), photons of a particular energy are *emitted*, and the spectrum of these photons is shown in Figure 2.3b. When white light is directed at a large number of sodium atoms, some of them will make transitions from a certain low energy level (E_{lo}) to a certain *higher* energy level (E_{hi}), and photons of a particular energy are *absorbed*. The spectrum of these photons is shown in Figure 2.3c.

Now, the numerical value of the energy of an atom that corresponds to each energy level (E_{hi} or E_{lo} in the example above) is not generally known. Rather it is the *differences* in energies between *pairs* of levels that are well defined, since it is these differences that correspond to the energies of the photons that the atom may absorb or emit. This is just the familiar principle of conservation of energy that you have met before, and can be summarised by saying that when an atom absorbs or emits a photon:

$$\text{photon energy} = \text{higher energy level} - \text{lower energy level}$$

or using symbols:

$$E_{ph} = \Delta E_{atom} \qquad (3.1)$$

where E_{ph} is the energy of the photon and ΔE_{atom} is the *change* in energy of the atom. (Remember from Book 3 that the Greek letter delta, Δ, is used to represent the *change* in a given quantity.)

- ■ What is the energy of the photons emitted or absorbed by the atoms when transitions take place between the energy levels E_{hi} and E_{lo} referred to above?
- □ Using Equation 3.1, the photon energy is equal to the higher energy level minus the lower energy level, so in this case the photon energy, $E_{ph} = E_{hi} - E_{lo}$.

In the case of sodium atoms, the two energy levels involved in the emission or absorption of the yellow line in the spectrum are separated by an energy difference of 2.1 eV. So the energy of the photons of yellow light in the sodium spectrum is 2.1 eV.

3.5 The energy levels of the hydrogen atom

You have seen that hydrogen atoms emit and absorb visible light of just five particular energies: 1.89 eV, 2.55 eV, 2.86 eV, 3.02 eV and 3.12 eV. It is important to remember that these are only the visible lines – hydrogen also emits and absorbs radiation that is not visible, for example in the infrared and ultraviolet regions of the electromagnetic spectrum. In order to build up a complete picture of the energy levels of the hydrogen atom, you would need a complete list of all the energies of the photons that it can emit or absorb.

Scientists have indeed produced such a list and it enables us to draw the complete **energy-level diagram** for the hydrogen atom (Figure 3.4a) in which the energy values are illustrated in a conventional and convenient way. What matters here is the vertical spacing of the energy values; neither the horizontal extent of the lines nor their thickness is of any significance whatsoever. One important thing to notice about this energy-level diagram is its simplicity: the pattern is one in which the energy levels gradually come closer together. As you can see from Figure 3.4a, the higher energy levels of hydrogen become so densely packed that it becomes impossible to draw them as distinct lines and the best that an illustrator can do is to indicate that the lines gradually get closer and closer together. The energy-level diagram for hydrogen is in fact the simplest of all atomic energy-level diagrams, just as you might have supposed.

You are now going to look a bit more closely at Figure 3.4a and think about it in terms of an analogy. You can imagine the energy levels in a hydrogen atom as being a little like the evenly spaced rungs of a ladder that is sunk into a deep pit (Figure 3.4b). In this analogy, the lowest rung of the ladder (the lowest energy level) is near the bottom of the pit, just above where the nucleus of the atom sits. When the hydrogen atom occupies its lowest energy level, you can imagine the electron sitting on this lowest rung, close to the nucleus. As the hydrogen atom occupies higher and higher energy levels, the electron sits on ever higher rungs of the ladder, until eventually the atom has so much energy that the electron climbs beyond the top rung and moves away from the pit. You will return to this analogy of a ladder in a pit later in this section.

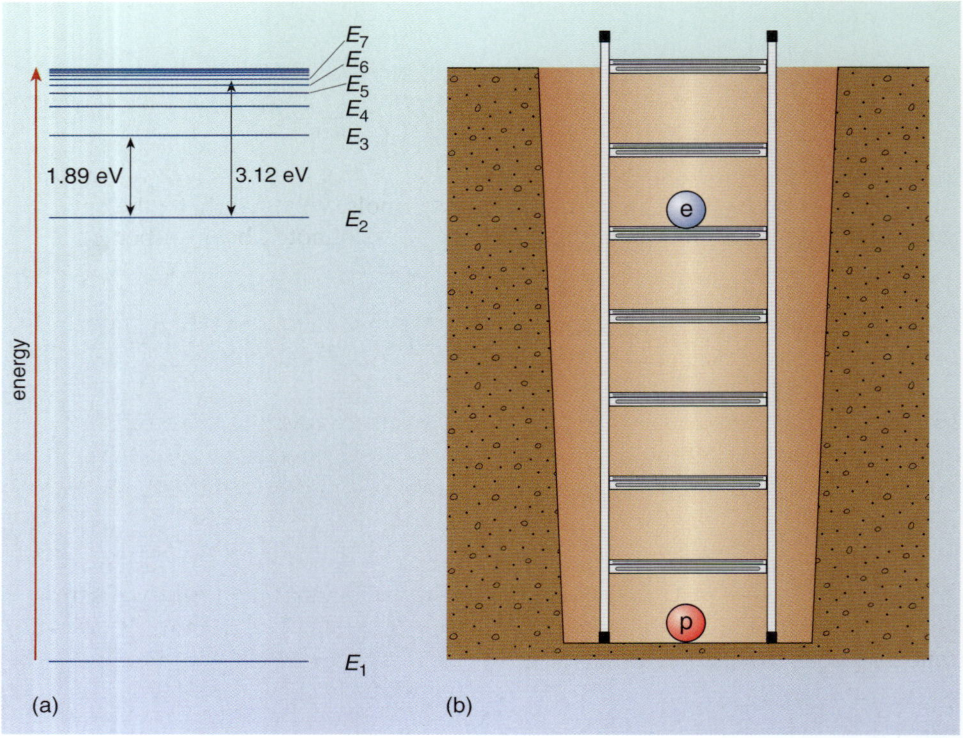

Figure 3.4 (a) The energy-level diagram of hydrogen. Energy increases moving up this diagram. The *differences* between energy levels represent the energies of photons that may be absorbed or emitted by hydrogen atoms. Two particular energy differences are shown, and the photons emitted or absorbed when hydrogen atoms make transitions between these energy levels correspond to the red line and one of the violet lines in the visible spectrum of hydrogen (Figure 2.4). (b) The energy-level diagram of a hydrogen atom can be thought of like the rungs of a ladder, sunk into a deep pit. You can imagine the hydrogen nucleus sitting at the bottom of the pit, and the electron sitting on one of the rungs. When the energy of the hydrogen atom corresponds to higher and higher energy levels, the electron can be thought of as sitting on ever higher rungs of the ladder. Broadly speaking, the higher the energy level, the greater the separation between the nucleus and the electron.

In Figure 3.4a the unknown energies of the various levels are identified only by the symbols E_1, E_2, E_3, etc., but the relative separations are drawn to scale, so that each of the photon energies associated with hydrogen is represented by the distance *between* two energy levels. For instance, photons of visible radiation (often called simply visible photons) with the *lowest* energy (1.89 eV) are associated with transitions between energy levels E_2 and E_3, whereas the visible photons with the *highest* energy (3.12 eV) are associated with transitions between E_2 and E_7. More widely separated levels indicate transitions of relatively large energy and so photons associated with these transitions may correspond to the ultraviolet part of the spectrum (energy differences greater than about 3.2 eV). Conversely, closely separated levels indicate transitions of relatively small energy and so photons associated with these transitions may correspond to the infrared part of the spectrum (energy differences less than about 1.8 eV). Remember, transitions do not just occur between adjacent energy levels, but between *any* pair of energy levels.

Question 3.1

(a) Suppose a hydrogen atom undergoes a transition that causes its energy level to change from E_3 to E_4. In terms of these symbols, what would be the energy of the associated photon? Would the photon be absorbed or emitted?

(b) Consider a transition that causes the energy level of a hydrogen atom to change from E_2 to E_1. In terms of these symbols, what would be the energy of the associated photon in this case? Would the photon be absorbed or emitted?

(c) A hydrogen atom makes a transition from E_7 to E_1. Bearing in mind the size of this jump when compared with those that produce visible photons, is the emitted photon in this case in the visible region of the spectrum?

(d) By absorbing a photon, a hydrogen atom makes a transition from E_5 to E_6. Bearing in mind the size of this jump when compared with those that produce visible photons, is the absorbed photon in this case in the visible region of the spectrum?

You will now look at the hydrogen energy-level diagram more closely. There is a lot to be learned from it and all the lessons apply to other atoms. Figure 3.5 repeats the pattern shown in Figure 3.4a, with some additional labelling.

When a hydrogen atom (or in fact any other atom) sits in the *lowest* energy level, labelled E_1, it is said to be in its **ground state**. (Remember this is like the bottom rung of the ladder in Figure 3.4b.) This is the 'normal' state in which an atom might be found. It will correspond to a situation in which the electron is likely to be close to the proton, with a minimum value of electrical energy. Above this energy, when

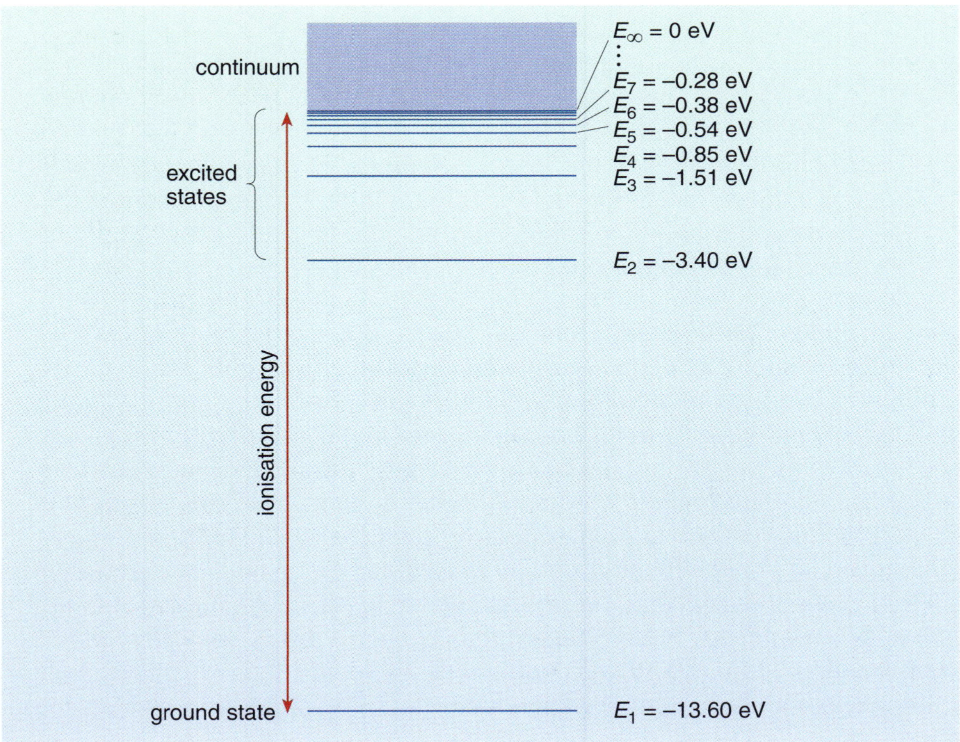

Figure 3.5 The energy-level diagram of hydrogen with some additional labels, which are explained in the text.

the atom has energy E_2, E_3, E_4 and so on, it is said to be in an **excited state**. (These states correspond to the higher rungs of the ladder in Figure 3.4b.) The topmost energy level is labelled E_∞, where the ∞ symbol is a mathematical shorthand that indicates 'infinity', the largest number imaginable. This topmost energy level corresponds to a situation in which the electron and proton are widely separated, with a maximum value of electrical energy. (This is when the electron is on the point of climbing out of the pit in the analogy of Figure 3.4b.)

The next point to note is that although Figure 3.5 indicates an infinite number of levels there is still only a finite difference in energy between the lowest energy level (E_1) and the highest energy level (E_∞). (The energy difference corresponds to the depth of the pit in the analogy of Figure 3.4b.) This shows that there is a limit to the amount of energy that a hydrogen atom can absorb without the electron and proton being split apart. If the atom is initially in its ground state, E_1, and it absorbs a photon with an energy greater than the difference ($E_\infty - E_1$), the electron is freed completely from its bondage to the proton at the core of the atom (the nucleus). In Book 4, you saw that the removal of an electron from a hydrogen atom leaves a charged particle which is written H^+ and is called an **ion**. (In fact the hydrogen ion is simply a proton.) The technical term for such removal of an electron from an atom is therefore **ionisation**. The energy difference ($E_\infty - E_1$) is called the **ionisation energy**, which in the case of hydrogen is measured to be 13.60 eV (to four significant figures).

There is a continuous range of possible energies above the ionisation energy and this is generally referred to as the **continuum**. It is shown in Figure 3.5 as a continuous 'band' stretching upward from E_∞. The continuum corresponds to situations in which the electron and proton move around separately, with a combined kinetic energy that is greater than E_∞. The electron is no longer bound to the proton, and the particles can have *any* value of energy.

■ A photon with energy 15.00 eV is absorbed by a hydrogen atom in its ground state. What happens to the atom?

□ The first 13.60 eV is used to ionise the atom, i.e. to separate completely the electron from the hydrogen nucleus, leaving 1.40 eV to be imparted to the liberated particles as kinetic energy. So, after the absorption, the result is that the (separated) electron and proton have a total kinetic energy of 1.40 eV.

Now for a rather subtle point. You have already seen that the energies of the levels are not really known. However, there is no such doubt about the *differences* in energy between the levels. These differences are very well defined, and, as you have already seen, they can be determined quite simply from atomic spectra. We don't really know the energy of any of the energy levels, but it makes life simpler for everyone if we all agree, as a matter of convention, that $E_\infty = 0$ eV. This corresponds to the energy of a separated electron and proton. (It is equivalent to the energy at the top of the pit in the analogy of Figure 3.4b.) All bound states of atoms, therefore, correspond to *negative* values of energy levels, as these energy levels sit below the '$E_\infty = 0$ eV' level. So, the ground-state energy is 13.60 eV below E_∞, namely $E_1 = -13.60$ eV. Furthermore, each of the excited levels can then be associated with some negative value of energy between −13.60 eV and 0 eV. This is the convention that is adopted for the rest of this book and it is also the convention that's been used to provide the energy values given on the right-hand side of Figure 3.5. Note that this is not implying that the atom has a 'negative energy' when it is in one of the bound states shown, simply that it has *less* energy than if the electron and proton were separated.

■ The spectral lines in the visible part of the hydrogen spectrum all correspond to transitions down to the E_2 energy level. From which higher energy levels must a hydrogen atom make a transition in order to produce photons of energy 1.89 eV, 2.55 eV, 2.86 eV, 3.02 eV and 3.12 eV?

□ To produce photons of energy 1.89 eV a hydrogen atom must make a transition from E_3 to E_2 since:

$$\begin{aligned} E_3 - E_2 &= (-1.51\ \text{eV}) - (-3.40\ \text{eV}) \\ &= (-1.51\ \text{eV}) + (3.40\ \text{eV}) = 1.89\ \text{eV}. \end{aligned}$$

(Remember that subtracting a negative number is the same as adding a positive number of the same size.)

Similarly:

$$E_4 - E_2 = (-0.85\ \text{eV}) - (-3.40\ \text{eV}) = 2.55\ \text{eV}$$

$$E_5 - E_2 = (-0.54\ \text{eV}) - (-3.40\ \text{eV}) = 2.86\ \text{eV}$$

$$E_6 - E_2 = (-0.38\ \text{eV}) - (-3.40\ \text{eV}) = 3.02\ \text{eV}$$

$$E_7 - E_2 = (-0.28\ \text{eV}) - (-3.40\ \text{eV}) = 3.12\ \text{eV}.$$

There is one final point to make about Figure 3.5. If you carefully examine the energies on the right-hand side, you will find that the energies of the various levels can be written as follows:

$$E_1 = -13.60\ \text{eV} = \frac{-13.60}{1}\ \text{eV}$$

$$E_2 = -3.40\ \text{eV} = \frac{-13.60}{4}\ \text{eV}$$

$$E_3 = -1.51\ \text{eV} = \frac{-13.60}{9}\ \text{eV}$$

$$E_4 = -0.85\ \text{eV} = \frac{-13.60}{16}\ \text{eV}$$

and so on.

■ Can you see a pattern in the numbers on the bottom of this series of fractions?

□ The number on the bottom is the square of the number that labels the energy level.

The energy E_n associated with the nth energy level is given by:

$$E_n = \frac{-13.60}{n^2} \text{ eV (where } n = 1, 2, 3, \text{ etc.)} \qquad (3.2)$$

This remarkably simple formula provides a wonderfully compact way of remembering everything that is shown in Figure 3.5! The formula applies only to the hydrogen atom; indeed, it is not possible to summarise all the energy levels of any other atom in a simple formula. Given the full range of energy levels represented in Figure 3.5, it is possible to identify *all* possible transitions between energy levels in the hydrogen atom and therefore all the possible spectral lines. In the case of the hydrogen atom these transitions are grouped into named series, some of which are shown in Figure 3.6.

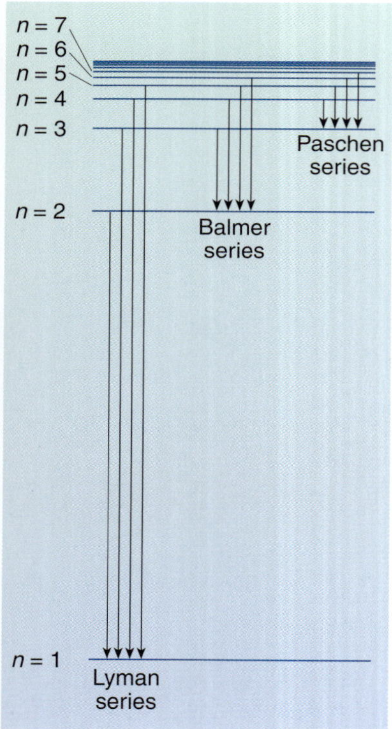

Figure 3.6 Some of the named series of transitions in the hydrogen atom that account for observed spectral series. All spectral lines that arise from transitions having the ground-state energy E_1 as their lowest level are said to belong to the Lyman series. Those with the $n = 2$ level as their lowest level are said to belong to the Balmer series, and those with lowest level $n = 3$ belong to the Paschen series. (These series are named after the scientists who discovered them.)

Question 3.2

(a) Use Figure 3.5 to calculate the energies of the first five lines in the Lyman series, in order of increasing energy.

(b) What is the energy of the photons corresponding to the highest-energy spectral line in the Lyman series?

(c) In which region of the electromagnetic spectrum would the Lyman series occur?

(d) What would this part of the spectrum look like if all the lines of the Lyman series were of equal brightness and you had the means to record them? Present your answer as a black and white sketch, and include an energy scale in electronvolts.

3.6 Quantum states and quantum numbers

As noted above, when the hydrogen atom is in its lowest energy level then it is in its ground state, and when it is in a higher energy level then it is in an excited state. A general name for these states of the atom is **quantum states** (or just *states* for short) and each quantum state has a precisely defined amount of energy – one of the energy levels of the atom. However, it turns out that there is *more than one* quantum state corresponding to each value of n in the hydrogen atom, and therefore associated with the nth energy level. The energy is virtually the same for all quantum states that correspond to a given value of n, but the possible values for the position and velocity of the electron in one state will be different from the possible values for the position and velocity of the electron in another state with the same value of n.

Before exploring these quantum states and how they may be described, here is one final piece of terminology: in the hydrogen atom, the *electron* can be said to *occupy* a particular quantum state of the atom. The reason for this is that it is only the possible values of the position and velocity of the *electron* that vary from one quantum state to another – the nucleus of the atom is unchanged. So when a hydrogen atom undergoes a transition from one quantum state to another, the position and velocity of the electron after the transition will be different from that before the transition. The crucial point is that energy levels and quantum states apply to atoms as a whole, but in the hydrogen atom, the (single) electron can be said to *occupy* a particular quantum state.

It turns out that there are 2 quantum states associated with the energy level E_1, 8 quantum states associated with the energy level E_2, 18 quantum states associated with the energy level E_3 and 32 quantum states associated with the energy level E_4. There is a pattern here.

> There are $2n^2$ different quantum states associated with the nth energy level of hydrogen.

To reiterate, each of the $2n^2$ quantum states associated with a given energy level will have *almost* the same energy, and for the purposes of this book it is generally sufficient to assume that they all do indeed have the same energy. However, the possible values for the position and velocity of the electron when it occupies each of those quantum states will be different from the possible values in another quantum state of the same energy.

■ How many different quantum states of hydrogen are associated with the energy level corresponding to $n = 5$? What makes them similar, and what makes them different?

□ There are $2 \times 5^2 = 50$ different quantum states associated with the $n = 5$ energy level. Each of these quantum states will have virtually the same energy, but different possibilities for the position and velocity of the electron.

If the number of quantum states corresponding to each energy level sounds familiar to you, it is probably because you're remembering about electron shells from Book 4 (Section 5.1). The present discussion is still about hydrogen atoms, but it turns out that the number of quantum states corresponding to each energy level applies to other atoms too, as you will see in Chapter 5. In fact, energy levels are essentially identical to what chemists refer to as *shells*, each of which can 'contain' a certain number of electrons. The number of electrons that a shell can hold is the same as the number of quantum states ($2n^2$) that has been worked out above.

The fact that there are different quantum states associated with a single energy level means that, to label a given state of a hydrogen atom, more is needed than just the number n that labelled the energy levels. Consequently, each of these different quantum states can be characterised by a unique set of **quantum numbers**. The number n is called the **principal quantum number** of the state and it can be any positive whole number (integer): 1 or 2 or 3 or 4, etc. To a good level of accuracy, the principal quantum number n determines the energy of any given quantum state in the hydrogen atom, according to Equation 3.2:

$$E_n = \frac{-13.60}{n^2}\,\text{eV} \tag{3.2}$$

All states with a particular value of n have virtually the *same* energy in the hydrogen atom, given by this formula. Also, as noted above, the principal quantum number labels the electron shell, when using that picture of how electrons are arranged in atoms.

To specify a state completely, however, the values of three more quantum numbers are generally needed. The first of these new quantum numbers is called the **orbital quantum number** and is usually given the symbol l. When used to

$l : 0 \ldots (n-1)$

$m_l : -l \ldots +l$

$m_s : \pm 1/2$

distinguish states that share a common value of n, it is only allowed to take the whole number values from 0 up to $n - 1$. For example, any state with $n = 1$ (the lowest allowed value of n) must have $l = 0$ (since $n - 1 = 0$), but among the eight different states with $n = 2$ there will be some with $l = 0$ and others with $l = 1$ (the maximum value of l since $n - 1 = 1$). Among the 18 states with $n = 3$ there will be some with $l = 0$, some with $l = 1$ and some with $l = 2$.

■ What are the possible values of l for states with $n = 5$?

□ Following the rule that l can take any value from 0 to $n - 1$, l could be 0, 1, 2, 3 or 4.

The two quantum numbers n and l still don't uniquely identify all the possible states corresponding to a particular energy level. The third quantum number is known as the **magnetic quantum number** and it can have any whole number value between $-l$ and $+l$. It is represented by the symbol m_l. So, when $l = 0$, $m_l = 0$ only. However when $l = 1$, $m_l = -1$, 0 or +1, and when $l = 2$, $m_l = -2, -1, 0, +1$ or +2, for instance. This gives a total of $(2l + 1)$ different values of m_l for any given value of l. Finally, the fourth quantum number m_s is known as the **spin quantum number** and it can be either $+\frac{1}{2}$ or $-\frac{1}{2}$ only.

■ How many different quantum states are there with $n = 5$ and $l = 3$?

□ When $l = 3$, m_l can take values of −3, −2, −1, 0, +1, +2 and +3 (that's seven different values). For each of these values, m_s can be either $+\frac{1}{2}$ or $-\frac{1}{2}$, making 14 different combinations of quantum numbers altogether. This corresponds to 14 different quantum states with $n = 5$ and $l = 3$.

The existence of $2n^2$ different quantum states corresponding to each value of n (and therefore to each energy level) can now be fully accounted for in terms of the rules that have been introduced, namely:

Rule 1 The principal quantum number, n, may be any positive whole number (1, 2, 3, etc.) and this number determines the energy level of the hydrogen atom.

Rule 2 For a given value of n, the orbital quantum number l may take any whole number value from 0 up to $n - 1$.

Rule 3 For given values of n and l, there are $(2l + 1)$ different values of the magnetic quantum number, m_l, ranging from $-l$ to $+l$.

Rule 4 For each combination of n, l and m_l there are two possible values of the spin quantum number, m_s.

As an example, consider the quantum states with energy E_2 that correspond to the first excited energy level of the hydrogen atom. All of these states will have $n = 2$ (Rule 1), so the only values of l they are allowed to have are 0 or 1 (Rule 2). How many of these states will there be in total? Well, using Rule 3, m_l will be 0 when $l = 0$ and it will be −1, 0 or +1 when $l = 1$. Finally from Rule 4, in each case m_s can be $+\frac{1}{2}$ or $-\frac{1}{2}$. Hence there are $2 \times ((2 \times 0) + 1) = 2$ states with $l = 0$, and $2 \times ((2 \times 1) + 1) = 6$ states with $l = 1$. The total number of possible states is therefore $2 + 6 = 8$. This is exactly the number predicted by the $2n^2$ rule you met earlier.

Question 3.3

Use the four rules above to demonstrate why there are $2n^2 = 18$ possible states with energy E_3.

It is rather tedious to keep saying 'a state with $n = 2$ and $l = 0$', to distinguish it from 'a state with $n = 2$ and $l = 1$', so a different convention is commonly used among scientists. This involves identifying the value of l by means of a letter, with 's' for 0, 'p' for 1, 'd' for 2, and 'f' for 3. The convention is to show the values n and l that correspond to a particular state by writing the numerical value of n and the letter representing l next to each other. Thus, the ground state of hydrogen, with $n = 1$ and $l = 0$, is a 1s state; an excited state of hydrogen with $n = 2$ and $l = 1$ simply becomes a 2p state; and a 4s state would be an excited state in which $n = 4$ and $l = 0$.

Figure 3.7 illustrates another way of thinking about quantum states and energy levels. The hydrogen energy-level diagram is shown yet again, but this time each of the energy levels characterised by the principal quantum number n is shown splitting into a set of quantum states characterised by the orbital quantum number l.

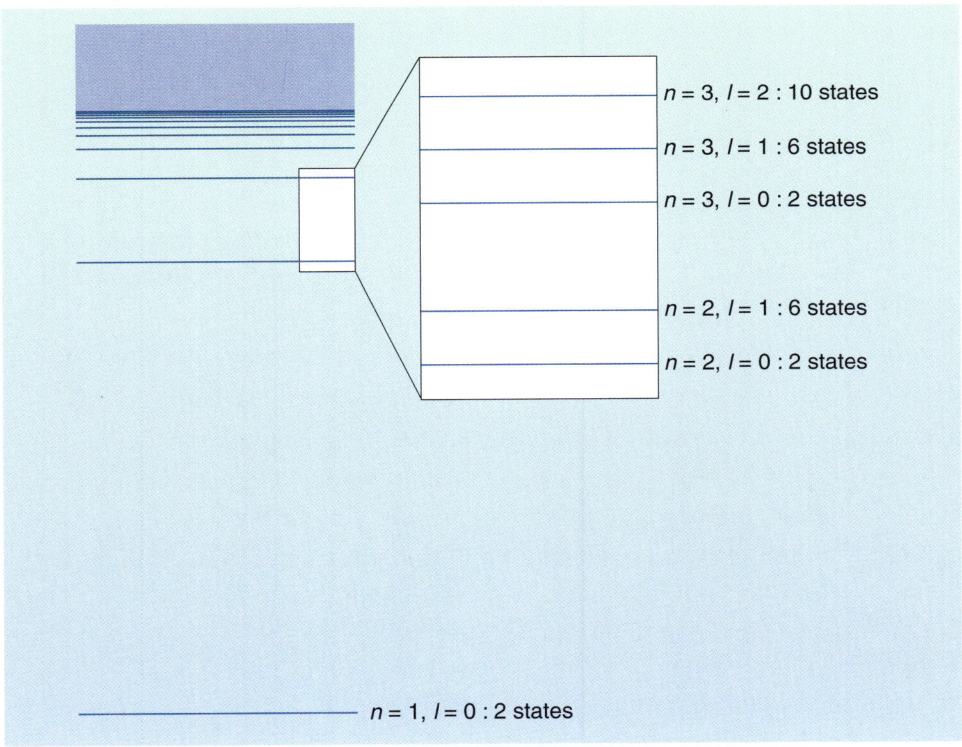

Figure 3.7 The hydrogen atom energy-level diagram showing how each energy level characterised by the principal quantum number n comprises multiple quantum states characterised by the orbital quantum number l. The range of energies of the levels corresponding to a given principal quantum number is exaggerated. The number of states corresponding to each combination of n and l is indicated.

Question 3.4

(a) A hydrogen atom in its ground state will sit in one of the two available 1s quantum states. What are the two possible sets of quantum numbers that the atom can have?

(b) When a hydrogen atom makes a transition between the two quantum states corresponding to the sets of quantum numbers in (a) it emits (or absorbs) a radio wave with a wavelength of 21 cm. What is the photon energy (in eV) corresponding to this radio wave, and therefore what is the separation of the two energy levels that comprise the 1s quantum state? (You will need to use Equations 2.3 and 2.4 and the following values: $h = 4.1 \times 10^{-15}$ eV s and $c = 3.0 \times 10^{8}$ m s^{-1} in order to answer this question.)

The 21 cm radio emission referred to in the question above has proved to be a vital tool in mapping the Galaxy. Most of the Universe is composed of hydrogen, and the structure of our own galaxy, the Milky Way, is delineated by the clouds of hydrogen gas that it contains. By tuning radio telescopes to pick up the 21 cm wavelength, astronomers can trace out where the hydrogen gas lies, and so unravel the structure of our local patch of space. It is worth remembering that this is only possible because the 1s energy level of the hydrogen atom is split into two distinct quantum states, separated by a few millionths of an electronvolt in energy.

The use of s, p, d and f to represent the numbers 0, 1, 2 and 3 arose for historical reasons. The letters once stood for 'sharp', 'principal', 'diffuse' and 'fundamental', from terms used to describe lines in spectra. But this doesn't really help you to remember them, and the use of the alpha-numeric convention is now so widespread that there is really no alternative but to commit it to memory. You should remember reading about *subshells* in Book 4 (Section 5.2), where you saw that chemists use these terms to refer to the way electrons are distributed in different atoms. Here then is the reason behind the subshell structure – electron subshells are simply quantum states that correspond to a particular set of quantum numbers. In a hydrogen atom, the single electron can occupy any one of the quantum states specified by the four quantum numbers that have been introduced. In other atoms, as you will see in the next chapter, the various electrons can be thought of as 'filling up' these quantum states in sequence.

Activity 3.1 Quantum numbers and quantum states

We expect this activity will take you approximately 15 minutes.

This activity allows you to consolidate your understanding of quantum states and quantum numbers in the hydrogen atom.

(a) Enter in the appropriate cells in Table 3.1 the notation for the quantum states of hydrogen that correspond to the allowed combination of quantum numbers with n between 1 and 4 and with l between 0 and 3. (Note that not all cells correspond to allowed combinations.)

(b) Enter in the table the values for the energy levels E_n in the hydrogen atom.

(c) Enter the total number of quantum states corresponding to each energy level in the right-hand column of the table, and enter the number of states corresponding to each particular value of the orbital quantum number in the final row of the table.

Table 3.1 Quantum states of hydrogen.

Principal quantum number, n	**Orbital quantum number, l**				**E_n/eV**	**Total number of quantum states for each n**
	0	1	2	3		
1	1s				−13.60	2
2	2s	2p			−3.40	8
3	3s	3p	3d		−1.51	18
4	4s	4p	4d	4f	−0.85	32
Number of quantum states for each l	2	6	10	14		

(d) Make a list of the set of four quantum numbers (n, l, m_l, m_s) that correspond to *each* of the allowed quantum states in the fourth energy level.

Now look at the comments on this activity at the end of this book.

3.7 Energy levels in general

The focus here has been on the energies of atoms and you have seen that all atoms have energy levels. This is the remarkable feature that is new and different about quantum physics. In the type of physics you met in Book 3, there was nothing to prevent an object from having *any* amount of kinetic energy, or gravitational potential energy, or electrical energy, or any other kind of energy. The energy of everyday objects, such as cars, can vary continuously and take any value you care to choose. However, down at the level of the quantum world of atoms, things are very different.

It is now time to look beyond atoms, to apply these ideas to other aspects of the quantum world. Quantum physics, in fact, makes a bold and very general prediction: whenever particles are bound together, they form something that has energy levels. Atoms are a familiar example of this – because every atom consists of electrons bound together with a nucleus, the atom will have energy levels.

■ Would you expect molecules to have energy levels? If so, what observable consequences would you expect?

□ Molecules are groups of atoms that are bound together, so you should expect that they will have energy levels. When molecules make transitions between their energy levels, they will emit or absorb photons which give rise to so-called molecular spectra.

■ Would you expect atomic nuclei to have energy levels? If so, what observable consequences would you expect?

□ Nuclei are groups of protons and neutrons that are bound together, so you should expect that they will have energy levels. When nuclei make transitions between their energy levels, they will emit or absorb photons which give rise to so-called nuclear spectra.

Experiments confirm these predictions of quantum physics; every day in laboratories all over the world, scientists study the spectra of molecules and nuclei. The idea of energy levels, so completely foreign to experiences in the everyday world, is a routine part of the quantum world. In summary:

> Every system in which particles are bound together, such as nuclei, atoms or molecules, will have *quantised* energy levels. Such systems are referred to as **quantum systems**.

■ Suppose you ionised a hydrogen atom, so that you freed its single electron from the atomic nucleus. Would you expect the separate electron and proton to be characterised by energy levels?

□ No – since the electron and proton are not bound together they do not have energy levels.

(a)

(b)

Figure 3.8 (a) The whitish glow of a tungsten filament electric light bulb and (b) the red glow of a hot-plate on an electric cooker. Both of these household appliances emit light that has a continuous spectrum. The glowing part of a hot-plate or of a filament lamp is composed of metal and the light coming from such objects is emitted by the metal as a whole, rather than from individual atoms.

After this discussion of the line spectra of atoms, you may be wondering how it is possible to get a continuous spectrum, such as that shown in Figure 2.2. Continuous spectra contain a continuous distribution of photon energies. They can be produced by switching on a conventional light bulb, which heats the tungsten filament to a very high temperature (Figure 3.8a), or by heating a plate on an electric cooker to a somewhat lower temperature (Figure 3.8b). The red glow from the hot-plate is not attributable to any particular photon energy; this spectrum is continuous, like that of the light bulb, except that the brightest part of the hot-plate's visible spectrum is the red part. Both these devices also emit radiation in other parts of the electromagnetic spectrum, such as infrared radiation.

The continuity of the spectrum from a heated object results from the fact that we are not observing emission from individual atoms, but the effect of many atoms together in a solid. In a solid metal, like tungsten, the atoms are arranged in a regular fashion, and some of the electrons are shared by the whole array of atoms. This is what makes the conduction of electricity possible. Though highly mobile, these electrons are confined, or bound, within the metal, so they are associated with energy levels. However, there are so many levels, and their energies are so close together, that they form an apparently continuous energy band that is typically a few electronvolts wide. Transitions within this band give rise to a continuous range of photon energies, and so produce a continuous spectrum. The energy levels of a metal, therefore, provide yet another example of quantised energy, in addition to the molecular, atomic and nuclear energy levels discussed above.

Question 3.5

Describe the general features that would be seen in the spectra produced by (a) a copper wire that is heated until it glows; (b) a vapour of copper atoms which is excited by an electric current being passed through it.

This chapter concludes by again considering the fundamental importance of quantum physics, this time by thinking about the technology that is enabled by our understanding of it. Since the 1920s, knowledge about quantum physics has continued to grow and diversify. It provides insights into a vast range of phenomena, and supplies the scientific basis for many 'high-tech' industries, ranging from the manufacture of computer chips to the fabrication of the tiny lasers that are used in DVD players. The age of information technology could never have come about without the underpinning science of quantum physics.

Fifty years ago, computers were rare objects, and they were vast in size owing to the bulky components from which they were built. Nowadays, there are computers all around us. Not just desktop or portable PCs, but mobile phones, televisions, car engines, and even washing machines each may contain computers (Figure 3.9). Modern computers are so compact because their circuitry comprises integrated circuits, sometimes also known as silicon chips. A miniaturised electronic circuit such as a computer CPU chip is only about 200 mm^2 in size, yet contains the equivalent of around a million transistors per square millimetre. At the heart of every miniature component lie semiconductor devices. These act as tiny electronic switches, controlling the flow of electric current according to the fundamental rules of quantum physics, which determine the arrangement of electrons in the semiconductor materials.

Figure 3.9 Examples of modern technology that rely on quantum physics.

Many people believe that the digital revolution brought about by the availability of integrated circuits was one of the most significant technological advances in the entire history of the human race, comparable to the development of the wheel, the printing press or the steam engine. It is worth remembering that it is our understanding of the underlying principles of quantum physics that allowed this revolution to occur.

3.8 Summary of Chapter 3

Quantum physics lies at the heart of understanding the behaviour of matter and radiation on the smallest scales. It therefore underlies not only the technology that surrounds us every day, but the fundamental structure and behaviour of everything in the Universe.

Each type of atom can be characterised by the energies of the photons it can absorb or emit. When these photons are dispersed to form a spectrum, the spectral lines of atomic spectra provide a unique 'fingerprint' of the atoms concerned. The explanation for atomic spectra is that atoms can only exist with certain values of energy, known as energy levels. Transitions, often referred to as quantum jumps, can occur between these energy levels. When an atom has its lowest possible energy, it is said to be in its ground state. To make atoms jump to excited states, which have a higher energy, photons of the correct energies must be supplied. When an atom jumps from one energy level to another of lower energy, the energy that it loses is taken away by a photon. In general, the energy of the photon is equal to the *change* in energy of the atom: $E_{ph} = \Delta E_{atom}$.

Hydrogen has a relatively simple energy-level diagram, and various series of transitions can be identified from this. The energy E_n associated with the nth energy level of hydrogen is given by:

$$E_n = \frac{-13.60}{n^2}\ \text{eV}$$

Consequently, the energy of the ground state ($n = 1$) is −13.60 eV.

There are $2n^2$ different quantum states corresponding to the nth energy level of hydrogen. Each state is uniquely identified by the values of a set of quantum numbers, the first two of which are denoted n and l. There are $2 \times (2l + 1)$ different states for each pair of allowed values of n and l, and l can take any whole number value from 0 up to $n - 1$. States with l values of 0, 1, 2 and 3 are described as s states, p states, d states, and f states respectively. In a hydrogen atom, all quantum states with the same value of n have essentially the same energy, but correspond to different possibilities for the position and velocity of the electron.

Any system of particles that are bound together will have quantised energy levels, but free particles will in general not have energy levels. Continuous spectra can be produced by hot objects in which there are many energy levels with extremely small separations between them. These levels form a continuous energy band within which transitions are possible.

Chapter 4
Quantum uncertainty

> If quantum physics hasn't profoundly shocked you, you haven't understood it yet.
>
> Niels Bohr, Danish physicist (1885–1962)

One of the great attractions of Newton's laws of motion that you met in Book 2 is that they deal in certainties. By using his laws of motion and gravity, everything in the Universe from the period of a pendulum's swing to the motion of the planets can be predicted with a complete and reassuring certainty. Yet scientists now know that it is simply not true that if you know the present state of something, you can necessarily predict its future with certainty. Never mind the entire Universe, it is not even possible to predict every aspect of the future of a single hydrogen atom!

You saw in the previous chapter how Planck's ideas shed light on atomic energies by showing that these energies have only certain, allowed values known as energy levels. But *why* do atoms have energy levels? This question could not be answered using the laws of physics that were known at the beginning of the 20th century. It became clear to a few young and exceptionally gifted scientists that a new approach was needed to describe the behaviour of atoms.

So, in the mid-1920s, quantum physics was born. In quantum physics we have to come to terms with intrinsic 'uncertainties' or, more accurately, *indeterminacies*. As you will see in this chapter, quantum physics says there are some things about atoms that are indeterminate; that is, they cannot be known or determined, no matter how clever you are or how much computer power you have to hand. For instance, it is impossible to say exactly *where* an electron in an atom is at a certain time and how *fast* it is moving or in which *direction*; the position and velocity are indeterminate.

Most people find it difficult to come to terms with the idea of quantum indeterminacy; Albert Einstein was the most famous conscientious objector to the very idea. He believed that it must be possible, if you have a good enough theory and good enough equipment, to probe, say, an atom in as much detail as you like. But, despite his brilliant arguments, nearly all quantum scientists believe that on this issue he was wrong. So be prepared to spend time getting to grips with **indeterminacy**, one of the most challenging concepts in quantum physics.

4.1 Indeterminacy and probability

When dealing with transitions between energy levels that require an increase in the energy of an atom, it is possible to exert some degree of control over which transitions occur by regulating the energies of the photons that are supplied to the atom. For example, suppose that a hydrogen atom is initially in its ground state, with energy E_1 on the energy-level diagram (the left-hand panel of Figure 4.1). You can ensure that the only upward transition this atom makes is to an energy

E_3 by only allowing the atom to interact with photons of energy $E_3 - E_1$. These photons carry just the right amount of energy to cause the desired transition, but too little to cause the jump to E_4, and too much to cause the jump to E_2. Even so, quantum physics allows the atom a degree of unpredictability in that although the atom *may* absorb a photon that has energy $E_3 - E_1$ it is not *required* to do so (the centre and right-hand panels of Figure 4.1). In a large number of identical encounters between hydrogen atoms and photons with this energy, all that can be predicted is the **probability** that the photons will be absorbed.

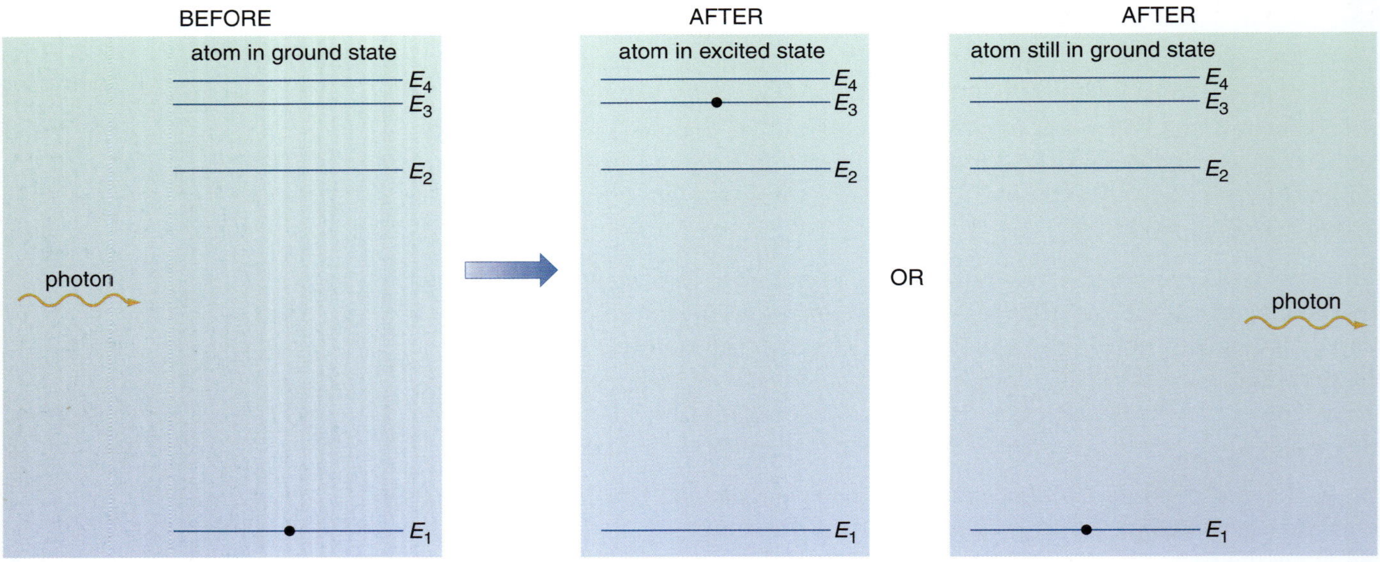

Figure 4.1 A photon with energy $E_3 - E_1$ may be absorbed by an atom in its ground state and excite the atom from energy level E_1 to energy level E_3, or it may just pass by leaving the atom unchanged. In quantum physics, it is not always the case that a photon that *can* be absorbed by an atom *will* actually be absorbed. Usually the best that can be done is to predict the probability of absorption.

The significance of probability in quantum physics is even more apparent if you consider downward, rather than upward, transitions. Once again consider a hydrogen atom, this time initially in the E_3 energy level (see Figure 4.2). Such an atom may spontaneously make a downward transition, giving out a photon in the process. If it does so, that transition may involve the relatively small jump down to the E_2 level and the emission of a 1.89 eV photon of red light (Figure 4.2 upper panel), or it might involve the much bigger jump down to the E_1 level and the emission of a 12.09 eV photon of ultraviolet radiation (Figure 4.2 lower panel).

Which transition will any particular atom make? The answer is that no one knows and, more significantly, according to quantum physics, *no one can know*. Given a large number of identical hydrogen atoms all in the same energy level, E_3, it is possible to predict the proportion of them that will emit ultraviolet photons and jump to E_1, and the corresponding proportion that will emit visible photons and jump to E_2, but it is *impossible* to predict which of the two possible jumps will be made by any particular atom. This idea can be summarised in the following way:

In quantum physics, the possible outcomes of a measurement can generally be predicted, and so can the probabilities of each of those possible outcomes. However, it is not generally possible to predict definitely the outcome of any individual measurement if there is more than one possible outcome.

It is believed that in quantum physics even the most completely detailed description of a system, such as an atom in a particular energy level, will still only allow predictions to be made about the probabilities of different outcomes for the future behaviour of the system. In other words, in quantum physics, the use of probability is an *essential* feature and not simply a matter of convenience or practicality.

The essential use of probability in quantum physics is the second of its defining characteristics.

Quantum physics, with the built-in imprecision of a description of nature based on probabilities, is supported by experiment. Not a single experiment has ever disproved quantum physics.

Question 4.1

In the light of what you have just read, which, if any, of the following statements would you say are correct?

(a) When a hydrogen atom in its ground state is isolated from other sources of energy, it has no possibility of change.

(b) Intense radiation, with photons of energies ranging up to 10 eV, has no effect on isolated hydrogen atoms in their ground state.

(c) If a hydrogen atom in its ground state absorbs a photon of energy 12.75 eV, there is no way of saying whether it will subsequently emit a visible photon.

You can see that in quantum physics there is a distinctive combination of strong prohibition and unpredictable freedom. Despite the remarkable prohibitions in items (a) and (b) of Question 4.1, the quantum world seems to allow great licence: whatever is not actually forbidden will be found to occur, sometime or other. In other words, what can happen will happen, sooner or later! None the less, the fact that even the most detailed possible description of a quantum system still involves probability in an essential way shows that indeterminacy is built into the heart of quantum physics.

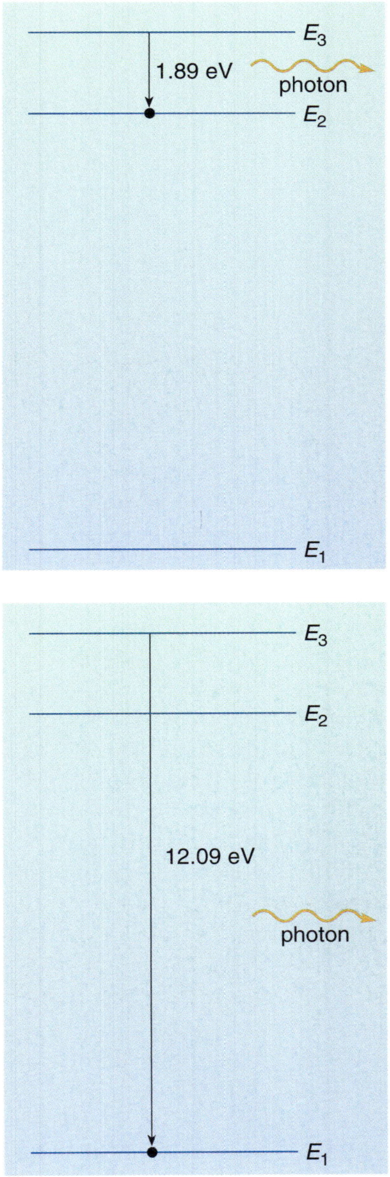

Figure 4.2 The two downward transitions available to a hydrogen atom in the E_3 energy level.

4.2 Modelling atoms

The structure of an atom cannot be seen directly, so instead scientists find it useful to come up with ways of describing atoms in terms of pictures or mathematical equations. Such a description is usually referred to as a **model**. (You initially met the concept of models in Book 1, particularly in the context of climate models.) If you have any preconceived picture of what an atom looks like, it may be that you imagine the electrons orbiting the nucleus, much as the planets in our Solar System orbit the Sun. Indeed, a model like this was proposed by the Danish physicist Niels Bohr in 1913. His model included fixed orbits for the electrons, each of which corresponds to one of the energy levels. Higher energy levels corresponded to orbits that were further from the nucleus (Figure 4.3a). Whilst this is a simple and appealing picture that can explain some phenomena, it has shortcomings that make it inconsistent with the real world. For instance, if electrons really did orbit the nucleus, then they would be constantly accelerating. (Remember from Book 2, Section 14.2 that an acceleration can correspond to a change in speed *or* a change in direction.) Any charged object undergoing an acceleration will continuously emit electromagnetic radiation and lose energy. As a result, an electron orbiting in this way would rapidly spiral into the nucleus as its electrical energy reduced. This does not happen, so electrons cannot really be orbiting the nuclei of atoms.

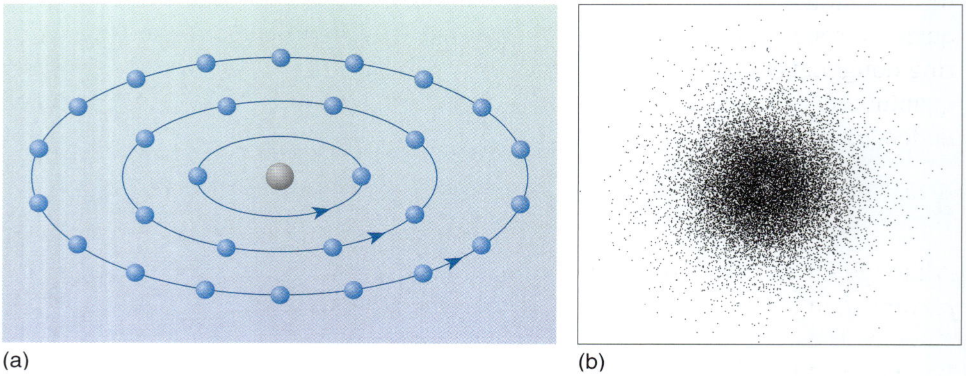

Figure 4.3 (a) The Bohr model of the atom with fixed orbits for the electrons corresponding to each energy level. (b) The Schrodinger model of the hydrogen atom says that the position of the electron is indeterminate, so it is represented here by a 'fuzzy cloud' surrounding the nucleus, indicating the range of possible positions that the electron may have. The cloud in this case is spherical, with the nucleus at its centre.

The richly detailed understanding of the chemistry and physics of atoms that modern science provides is firmly rooted in a different model of the atom that fully incorporates the essential indeterminacy of quantum physics. The quantum model of the atom that is used today emerged from the work of a number of European physicists in the mid-1920s. One of the most influential members of this group was the Austrian theoretical physicist Erwin Schrödinger (1887–1961), so it is convenient to refer to the model as the **Schrödinger model of the atom**. The model is mathematically complex, but here the focus is only on its basic concepts and results rather than the methods used to obtain them. As usual the discussion concentrates on the hydrogen atom, but it is worth noting that the model can successfully describe all other atoms too. When the Schrödinger model is applied to the hydrogen atom, it predicts that the atom has energy levels given by Equation 3.2. It also predicts the existence of each of the $2n^2$ quantum states associated with each energy level, and the allowed values of quantum numbers in each case. Finally, Schrödinger's model predicts detailed information about the possible position and velocity of the electron when the atom sits in each of these quantum states.

The first of these points confirms that the Schrödinger model accounts for the emission and absorption spectra (the 'spectral fingerprint') of hydrogen. The final point predicts that the behaviour of the electron in the atom is intrinsically indeterminate. According to quantum physics, if a hydrogen atom sits in a particular quantum state (with a certain value of energy), it is impossible to say in advance of a measurement what values will be obtained for the electron's position or velocity. This means that if *identical measurements* are made on atoms in the *same* quantum state, the experiments will give a *variety of different outcomes*. What *can* be predicted are all the possible outcomes of a measurement (i.e. all the possible positions and velocities of the electron) and the probabilities of each of those outcomes.

It turns out that the most probable position for an electron in a hydrogen atom is further from the nucleus for the higher energy levels, and closer to the nucleus for the lower energy levels. However, a large range of positions for the electron are possible in each energy level. To draw an image of the atom, the best that can be done is something like Figure 4.3b where the 'fuzzy cloud' surrounding the nucleus represents all the possible positions that the electron may have. Different quantum states correspond to different sets of possible values for the electron's position, and so will be described by different electron clouds. Some of these clouds are spherical (like that shown in Figure 4.3b), whereas for some quantum states, the electron clouds are shaped like dumb-bells, clover leaves or ring doughnuts! Here then is a further demonstration of indeterminacy at work in quantum physics and another illustration of the essential use of probability in our understanding of how atoms behave.

4.3 The uncertainty principle

Although the precise position and velocity of an electron in an atom cannot be *predicted*, in theory it is possible to devise experiments to *measure* its position or velocity. But can *both* the position and velocity be measured simultaneously, so tying down the behaviour of the electron precisely? It turns out that the answer is no. A fundamental result of quantum physics, discovered in 1927 by Werner Heisenberg (1901–1976), explicitly rules out such detailed knowledge.

> The Heisenberg **uncertainty principle** rules out the possibility of combining definite knowledge of some quantities (such as position) with definite knowledge of certain other quantities (such as velocity). It also limits the precision with which such quantities can be measured simultaneously.

According to the uncertainty principle, measuring one quantity with a prescribed level of precision automatically limits the precision with which the other can be known at the time of the measurement. Indeed, the uncertainty principle says that if you know the exact position of a particle, you can know nothing at all about its velocity at the same time. The converse is also true: if you know its velocity exactly, you can't know anything about its position simultaneously.

To get some idea of why this is the case, think about what measuring the position of an electron might actually involve. The obvious way to determine the position is to 'see' where the electron is, and to 'see' where something is you have to shine light on it. But shining light on an atom will have one of two effects. On the one hand, a photon may be absorbed, so causing the atom to jump to another energy level, in which case the quantum state of the atom has changed. Alternatively, the photons may not interact with the atom, but emerge unaffected, in which case you haven't measured anything! In other words, by measuring the position of the electron in an atom, you will change the energy of the atom, and so alter the possible range of velocities of the electron. Indeed, *any* act of measurement on a hydrogen atom, or any other quantum system, will involve transferring energy into or out of that system, and so will change its properties.

Question 4.2

Which of the following statements are correct? Explain your answer in each case.

(a) Heisenberg's uncertainty principle says that everything about the motion of a particle is uncertain.

(b) The more precisely the velocity of a particle is known, the less precisely can its position be known at the same time.

(c) Eventually, an experimenter is bound to build apparatus that will enable both the position and velocity of an electron in an atom to be determined to arbitrarily high precision.

(d) When he formulated his uncertainty principle, Heisenberg was implicitly criticising the skill of his experimental colleagues.

It is important to appreciate that the limitations of the uncertainty principle are a matter of deep principle in quantum physics, not a result of sloppy work or poor equipment. The quantum world is not only essentially indeterminate, it is also inherently uncertain.

Perhaps the ultimate expression of quantum indeterminacy is the fact that the smallest constituents of matter, such as electrons, really do not behave like particles, localised at a point in space, at all. As you read in Book 3, Section 11.2, the French physicist Louis de Broglie (1892–1987) proposed that, just as electromagnetic radiation propagates like a wave and interacts like a particle, so *any* moving 'particle' has an associated wave, and will therefore exhibit wave-like behaviour such as diffraction by a suitable grating. The wavelength associated with any particle depends on its mass and speed – the smaller the mass and the slower it moves, the larger is its **de Broglie wavelength**. For instance, the de Broglie wavelength of an electron moving at 1% of the speed of light is about 0.2 nm, and a beam of electrons will therefore be *diffracted* by the regular lattice formed by atoms in a crystal, whose spacing is about 0.5 nm.

De Broglie's idea of **wave–particle duality** may be summed up by noting that electrons interact like particles, but propagate like waves.

It is a remarkable irony that the English physicist J. J. Thomson (1856–1940) had been awarded the Nobel Prize for his discovery of the electron, displaying properties of a particle, in 1897, and that his son, George Thomson (1892–1975), also received the Nobel Prize for his demonstration of electron diffraction, proving that electrons can behave like waves, in 1927. Both models of the electron's behaviour are appropriate in different circumstances.

In the everyday world, however, we can largely forget about quantum indeterminacy, the Heisenberg uncertainty principle and de Broglie's wave–particle duality. The reason for this is that atoms, and other quantum systems, are so very small and have such a tiny mass. The more massive an object, the less important are the effects of indeterminacy. For instance, the uncertainties in the position or velocity of a planet orbiting the Sun are so tiny that they are utterly insignificant and, in effect, its position and velocity can be measured as precisely as measuring instruments allow. Likewise, the de Broglie wavelength of, say, a person is so tiny that you are not noticeably diffracted when you walk through a doorway! In general, quantum indeterminacy is only apparent when dealing with things on an atomic scale, or smaller. But it is this very reason why an understanding of quantum effects is necessary to construct the integrated circuits of semiconductor devices that lie at the heart of the technological world. Without knowledge of quantum indeterminacy, the uncertainty principle and wave–particle duality, our world would be a very different place.

4.4 Summary of Chapter 4

In quantum physics, the possible outcomes of a measurement can generally be predicted, and so can the probabilities of each of those possible outcomes. However, it is not generally possible to definitely predict the outcome of any individual measurement when there is more than one possible outcome.

The Schrödinger model of the atom embraces the indeterminacy of quantum physics. In the case of the hydrogen atom, the model provides definite predictions for the energy levels, and probabilistic predictions concerning the position and velocity of the electron. It is useful to represent the possible positions of electrons in atoms by 'fuzzy clouds' surrounding the nucleus. Different clouds will correspond to each quantum state.

The Heisenberg uncertainty principle states that one cannot simultaneously have precise knowledge of the position and velocity of an electron (or any other quantum system). It emphasises the impossibility of avoiding indeterminacy in the quantum world.

Particles, such as electrons, can be characterised by their de Broglie wavelength. The fact that electrons interact like particles but propagate like waves lies at the heart of wave–particle duality, and underlies the computer technology on which the modern world relies.

Chapter 5
Atomic structure

> Nothing exists except atoms and empty space; everything else is opinion.
>
> Democritus, Greek philosopher (460–370 BC)

Having examined some of the distinctive characteristics of quantum physics, you are now ready to apply these principles to learn about the structure of atoms, the processes in which their nuclei can participate, and the fundamental particles of which they are composed. As you work through Chapters 5, 6 and 7 you will see the ideas that have been developed in Chapters 3 and 4 being applied on smaller and smaller length scales. The evidence indicates that the principles of quantum physics, and the characteristic quantum phenomena of energy levels and indeterminacy, continue to provide a reliable guide to the nature of the physical world on all these length scales.

5.1 Atoms and ions with a single electron

The aim of this chapter is to develop an understanding of the structure of atoms in general. A start has already been made on this, in Chapters 3 and 4, by examining the hydrogen atom in particular. Knowledge of the quantum behaviour of the one electron that is found in a hydrogen atom will provide a guide towards understanding the behaviour of the many electrons that are found in more complicated atoms. Of course, you should not expect to understand the full richness of atomic behaviour on the basis of a brief introduction to the hydrogen atom, but what you already know can be used to provide an insight into the behaviour of helium and lithium atoms (the two simplest atoms after hydrogen), and those atoms illustrate many of the general principles that govern the structure of even more complicated atoms. This in turn is the basis for the chemical behaviour of atoms that you have read about in Book 4, which relies crucially on the way that the electrons are arranged around the nucleus in each different element.

Activity 5.1 Electrons in atoms

We expect this activity will take you approximately 90 minutes.

The aim of this computer-based activity is to reinforce your understanding of some of the important concepts of quantum physics that you met in Chapters 3 and 4, particularly the relationships between energy levels, quantum states and spectra. You will initially explore these relationships by investigating the hydrogen atom, and you will then discover the differences between the energy levels and spectra of hydrogen atoms, and those of singly ionised helium ions, He^+, and doubly ionised lithium ions, Li^{2+}.

You should now work through the computer-based activity *Electrons in Atoms*. When you have completed this, look at the comments at the end of this book.

■ What is it that makes the hydrogen atom, the He^+ ion and the Li^{2+} ion similar?

□ All of these have the same number of electrons, in this case just a single bound electron, and can therefore be described as **isoelectronic**. He^+ ions and Li^{2+} ions are referred to as **hydrogen-like ions**.

In Activity 5.1 you saw that the energy levels of a hydrogen-like ion with nuclear charge Ze are given by:

$$E_n = Z^2 \times \frac{-13.60}{n^2}\ \text{eV} \tag{5.1}$$

which differs from Equation 3.2 (the corresponding equation for the hydrogen atom) only by the factor of Z^2. Since $Z = 1$ for the hydrogen atom, the two equations are identical in that case, so Equation 5.1 is really just a generalisation of Equation 3.2.

As you saw in Activity 5.1, in the case of the helium ion, He^+, $Z = 2$, so Equation 5.1 implies that each value of E_n is four (i.e. Z^2) times larger than the corresponding value for the hydrogen atom. It also follows that the separation between any two energy levels in the He^+ ion will be four times greater than the separation between the corresponding levels in the hydrogen atom.

Question 5.1

(a) The Be^{3+} ion, with atomic number $Z = 4$, has one bound electron. Use Equation 5.1 to calculate the energy corresponding to the ground state of this ion.

(b) The energy of the first excited state (i.e. $n = 2$) of the Be^{3+} ion turns out to be very close to the energy of the ground state of the one-electron helium ion, He^+. Why is this?

Question 5.2

When a hydrogen atom makes a transition from a state with principal quantum number $n = 3$ to a state of lower energy, with $n = 2$, a photon of light carrying an energy of 1.89 eV is emitted, contributing to the red line in the spectrum.

(a) What is the corresponding photon energy for this jump in the helium ion, He^+?

(b) A transition from the $n = 6$ to the $n = 4$ energy level in He^+ is accompanied by the emission of an identical photon to that emitted when a hydrogen atom makes a transition from the $n = 3$ to $n = 2$ energy level. Using Equation 5.1, explain why this is so.

5.2 Atoms and ions with two electrons

There is *no* simple formula like Equation 3.2 for the energy levels of any neutral atom other than hydrogen. Quantum physics can predict the energy levels of the helium atom (for which $Z = 2$) to high precision, but it takes a lot of computer time to achieve this. Generally speaking, it takes even more computer time to determine the energy levels for atoms with three or more electrons.

Another point to note is that, for hydrogen-like atoms and ions, the energy of each state is essentially determined by just the principal quantum number n. All 2s and 2p excited states for instance have virtually the same energy level in hydrogen, as noted in Chapter 3. In atoms with two or more electrons, things are no longer that simple. Interactions between the electrons mean that the energy levels of each state depend on the other quantum numbers, such as l, as well as on n.

The reason for this increasing complexity is not particularly related to quantum physics. Similar complications would arise in attempts to use Newton's laws to predict the behaviour of the Solar System if it contained two planets that attracted each other as strongly as each is attracted to the Sun. In the case of the helium atom the analogous problem is to find the positions and velocities for two electrons, in states of definite energy, when each electron is subject to the attractive influence of the nucleus *and* to the repulsive influence of the other electron. Just in case this does not already sound challenging enough, bear in mind that this is quantum physics, so the positions and velocities of both electrons will be indeterminate!

In the case of the hydrogen-like ions, we were quite prescriptive about the quantum state that's occupied by the single bound electron. If the ion is in its ground state, the electron occupies a 1s state; if the ion is in an excited state, then the electron might occupy a 2s state, or a 2p state, or any of the excited states that were described in Chapter 3. These states will not be exactly like those of the hydrogen atom, but for these simple ions the electrons can be fitted into 'hydrogen-like' states since the only thing that changes is the charge on the nucleus.

Now, moving on to consider atoms or ions with two bound electrons, such as He or Li^+, the state of the atom can once again be described in terms of the behaviour of the electrons. For instance, an atom in a quantum state described by 1s1s would indicate that it contains two electrons each occupying something like a 1s state of a hydrogen-like atom. An atom in a quantum state described by 1s2s would indicate two electrons, with one of them occupying a 1s hydrogen-like state and the other a 2s hydrogen-like state. But what are the energy levels corresponding to these quantum states?

Consider first the energy-level diagram for the helium atom, which has two electrons. The ground state of helium will be described by 1s1s and so will correspond to two bound electrons each occupying something like a 1s state of a hydrogen-like atom. To depict the energy-level diagram, the first thing to decide is where to position the zero of energy. In the case of the hydrogen atom, the zero was chosen to correspond to a state in which the electron just manages to escape from the nucleus. Figure 5.1 shows the energy-level diagram for the helium atom, and, as you can see, the zero energy level has been chosen to correspond to a helium nucleus and two free electrons. The energy level corresponding to the ground state of singly ionised helium He^+ is also shown, with the value (−54.40 eV) that you saw in Activity 5.1.

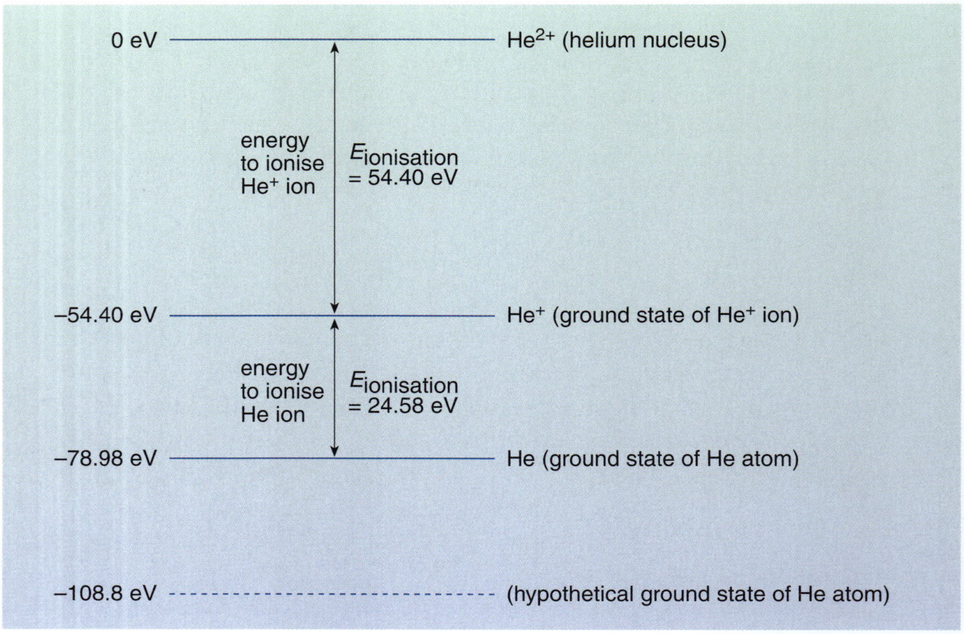

Figure 5.1 The energy-level diagram for the helium atom. This shows energy levels for the helium nucleus He^{2+} plus two free electrons, for the ground state of the He^+ ion plus one free electron, and for the ground state of the helium atom He.

Now suppose, for a moment, that the repulsion of the two electrons in the ground state of the helium atom could be ignored. Then the ground-state energy would correspond to assigning principal quantum number $n = 1$ to each electron, and adding the energies associated with each.

■ Where would such a hypothetical ground state be?

□ The ground state of the He^+ ion is at −54.40 eV, so the hypothetical ground state of a helium atom with two bound electrons might be expected to be at an energy of $2 \times (-54.40\ \text{eV}) = -108.8\ \text{eV}$.

Figure 5.1 shows that, in reality, the true ground state of the helium atom is significantly higher, at an energy of −78.98 eV. To understand, qualitatively, why the true ground-state energy of Figure 5.1 is higher than the hypothetical ground state with no interaction, think about where the two electrons may be found. The positions are indeterminate, but there is some probability of finding one electron far from the nucleus and one electron much closer to it. In this case, the distant electron effectively experiences only the electric force due to the He^+ ion, with charge $+e$ (i.e. the nucleus plus one electron). Conversely, the electron that is much closer to the nucleus experiences the effect of the full nuclear charge, $+2e$. This being the quantum world, all possibilities between these two extremes are allowed, and the typical charge experienced by each electron will be something between $+e$ and $+2e$. The name given to this effect is *screening* and you can think of it as one electron neutralising, or cancelling out, part of the charge of the nucleus. You saw in Activity 5.1 that the energies of the quantum states depended strongly on the nuclear charge. So it should come as no surprise that, when one electron reduces the effective charge experienced by the other, the energy levels will change dramatically.

In the general case of a two-electron ion, with nuclear charge Ze, the 'effective' nuclear charge, determining the ground-state energy, lies somewhere between $(Z-1)e$, corresponding to complete screening of one unit of charge by an inner electron, and Ze, corresponding to no screening.

Equation 5.1 gives the ground-state energy for any ion with just a single electron as $E_1 = Z^2 \times (-13.60\text{ eV})$. A first estimate for the ground-state energy (E_{est}) of a two-electron ion might be just twice this value, $E_{\text{est}} = 2 \times Z^2 \times (-13.60\text{ eV})$. But as you have seen above, the effective charge experienced by each electron is not Ze but something between $(Z-1)e$ and Ze. Let's call this effective charge $Z_{\text{est}}e$. So a better estimate for the ground-state energy of a two-electron ion is:

$$E_{\text{est}} = 2 \times Z_{\text{est}}^2 \times (-13.60\text{ eV}) \qquad (5.2)$$

The Schrödinger model for an atom or ion with two electrons gives rise to a detailed and complicated equation for its energy levels. By making approximations to the difficult maths required to solve this equation, the value of Z_{est} that determines the effective nuclear charge experienced by two electrons bound to a nucleus of charge Ze is found to be:

$$Z_{\text{est}} = Z - 0.3125 \qquad (5.3)$$

To see how the estimate fares, set $Z = 2$, for the helium atom. In this case, $Z_{\text{est}} = 2 - 0.3125 = 1.6875$ and therefore:

$$E_{\text{est}} = 2 \times (1.6875)^2 \times (-13.60\text{ eV}) = -77.46\text{ eV}$$

Comparing this with the experimental result of -78.98 eV, you can see that the actual value lies only about 1.5 eV below the estimate.

You have seen that the ground-state energy of the helium *atom* is -78.98 eV and that the ground-state energy of the helium *ion* is -54.40 eV. In order to turn an atom into an ion, the atom must be ionised. So the ionisation energy of the helium atom is the energy required to remove one electron from the atom in its ground state. As shown on Figure 5.1, the energy required to do this corresponds to the energy of the helium ion ground state minus the energy of the helium atom ground state:

$$E_{\text{ionisation}} = -54.40\text{ eV} - (-78.98\text{ eV}) = (78.98 - 54.40)\text{ eV} = 24.58\text{ eV}$$

This turns out to be the *biggest* ionisation energy for any neutral atom.

You are now in a position to make quantitative predictions about the ground-state energy and ionisation energy of the singly ionised lithium ion, Li^+. This has two bound electrons and is therefore isoelectronic with the helium atom. It too has a ground state denoted by 1s1s, the only difference is its atomic number, $Z = 3$.

Question 5.3

(a) Use Equations 5.2 and 5.3 to obtain an estimate for the ground-state energy of the Li^+ ion.

(b) Use Equation 5.1 to calculate the ground-state energy of the Li^{2+} ion.

(c) Using the two numbers you have just calculated, what is the energy required in order to remove an electron from the ground state of the Li^+ ion and so turn it into a Li^{2+} ion?

5.3 Atoms with three or more electrons

As you have just seen, it takes about 122 eV to remove the electron from a Li^{2+} ion, and it takes about 74 eV to remove one of the electrons from a Li^+ ion. However, it turns out that only about 5 eV is required to remove one of the electrons from a neutral Li atom. No singly ionised ion requires more energy than Li^+ to make it doubly ionised, yet the lithium atom itself is rather easily ionised, much more easily than the helium atom or the hydrogen atom.

The key to understanding why is to consider what happens when a third electron is added to the ground state of Li^+, to make a neutral lithium atom. As you saw in Chapter 3, there are only two quantum states that have quantum numbers $n = 1$ and $l = 0$, i.e. there are only two 1s states, each of them has $m_l = 0$, one of them has $m_s = +\frac{1}{2}$ and the other has $m_s = -\frac{1}{2}$. You can think of these hydrogen-like states as though they can each be 'occupied' by a single electron. So the ground state of lithium cannot correspond to a quantum state of 1s1s1s.

The principle that bans the third electron from being in a similar state to the other two is a crucial result of quantum physics; it was suggested by Wolfgang Pauli (1900–1958).

The Pauli **exclusion principle** bans any two electrons in the same atom from occupying the same quantum state.

Remember from Chapter 3 that for any value of n, there are two 's' states, six 'p' states, ten 'd' states, and so on. Each quantum state corresponds to a different allowed combination of the four quantum numbers. According to the exclusion principle, each of these quantum states can accommodate only one electron.

So, since there are only two 1s states, the third electron in a lithium atom must occupy a 2s state. The ground state of the lithium atom can therefore be represented as 1s1s2s. The third electron in the lithium atom has principal quantum number $n = 2$, which makes it much more remote from the nucleus. So, one of the three electrons in the lithium atom is rather weakly bound, since it experiences a net charge that is not much greater than the one unit of the Li^+ ion, because the other two units of nuclear charge are effectively screened by the other two more tightly bound electrons. A rough estimate of the ionisation energy of the lithium atom would be somewhat greater than that for the 2s or 2p state of hydrogen (i.e. the E_2 energy level), namely:

$$\frac{13.60\ \text{eV}}{2^2} = 3.40\ \text{eV}$$

In fact it is 5.39 eV, indicating that the other two electrons do not completely screen two units of nuclear charge.

As you have seen, for simple atoms such as He and Li, we are fairly prescriptive about the states occupied by the two or three electrons that now have to be considered. For the helium atom, both electrons can occupy 1s states, so the ground state of helium is 1s1s. For the lithium atom, you have seen that the ground state is 1s1s2s.

With more electrons, this notation soon gets unwieldy. To avoid this problem, a more compact way of describing the organisation of the electrons around a nucleus is used, referred to as the **electron configuration**. In fact, you have already met this notation in Book 4 (Section 5.2), where it was introduced as the way in which chemists refer to the arrangement of electrons in atoms. For the helium atom which had a ground state of 1s1s, the electron configuration is written as $1s^2$, and for the lithium atom which had a ground state of 1s1s2s, the electron configuration is written as $1s^22s^1$. It is important to realise that the superscripts here *do not* refer to 'powers' of numbers, they are merely labels indicating the number of electrons that occupy the states specified by the n and l quantum numbers.

As you saw in Book 4, it is often quite convenient to think of electrons as 'filling up' successive subshells in the atom. For instance: the 1s subshell can accommodate two electrons since there are two 1s quantum states, the 2s subshell can accommodate two more electrons, and the 2p subshell can accommodate a further six electrons since there are six 2p quantum states. Subshells are filled in order of increasing energy so that the atom as a whole has the lowest possible energy level. Although such terminology is not entirely accurate, owing to the interactions between electrons, it provides a useful simplification. For atoms that contain several electrons the states are somewhat different from the 'hydrogen-like' quantum states, because the interactions between the electrons complicate matters.

Question 5.4

(a) What do you suppose is the ground state electron configuration for a beryllium atom with four electrons?

(b) What is the ground state electron configuration for a boron atom with five electrons?

(c) What is the largest number of electrons that can be accommodated by filling the 1s, 2s, and 2p subshells?

You're now in a position to appreciate how and why the pattern of electronic configurations and subshell structure arises for the rest of the elements that you saw in Book 4. All the richness of chemistry and chemical reactions essentially stems from these simple quantum rules for how the electrons are distributed in atoms. This chapter concludes by emphasising the remarkable fact that lithium is a highly reactive metal, but helium is an extremely inert gas (Figure 5.2). Yet the only difference between them is that lithium atoms contain three electrons surrounding a nucleus containing three protons, whereas helium atoms have two electrons surrounding a nucleus with two protons. The difference in properties is all due to the ionisation energy difference between the two atoms, which in turn is due to the fact that quantum physics and the Pauli exclusion principle place a limit on the number of electrons in different quantum states.

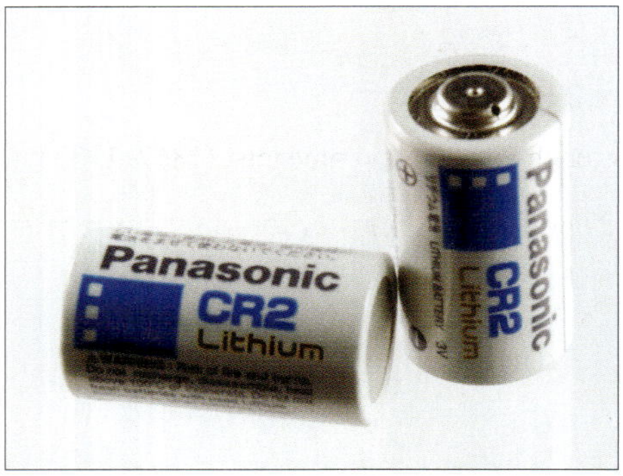

(a)

(b)

Figure 5.2 Lithium is a highly reactive metal, but helium is an extremely inert gas.

5.4 Summary of Chapter 5

Hydrogen-like ions have energy levels given by:

$$E_n = Z^2 \times \frac{-13.60}{n^2}\ \text{eV}$$

As a consequence of this, the energy levels of He^+ ions are further apart than those of hydrogen atoms by a factor of four and the photon energies in the spectrum of He^+ ions are greater than those of hydrogen atoms by a factor of four.

In atoms and ions with two electrons, one electron will partially screen the charge on the nucleus from the other electron. This raises the energy levels above what would be predicted if there was no interaction between the electrons.

The Pauli exclusion principle bans any two electrons in the same atom from occupying the same quantum state. It plays a vital role in determining the way electrons are arranged in atoms, and therefore underlies their chemical behaviour. One consequence of the exclusion principle is that any 's' state can only accommodate two electrons. Lithium, with three electrons, therefore has the ground state electron configuration $1s^2 2s^1$, making it much easier to ionise than helium with the ground state electron configuration $1s^2$.

It is often convenient to think of electrons as filling up successive subshells in an atom. Each subshell is characterised by a particular combination of n and l quantum numbers, and is only able to accommodate a certain number of electrons, equal to the number of quantum states with that particular combination of n and l.

Chapter 6 Nuclear processes

> Nothing in life is to be feared, it is only to be understood. Now is the time to understand more, so that we may fear less.
>
> Marie Curie, Polish-French chemist and physicist (1867–1934)

You are now going to travel further inside the atom to look at the quantum world of nuclei. You know already that nuclei have two building blocks: neutrons and protons, and these are referred to collectively as nucleons. It is simply different combinations of nucleons that account for the different isotopes in the world around us.

Even though you will be entering a new realm, you can be sure that there are many similarities between the quantum world of atoms and electrons that you have just read about, and the quantum world of nuclei. For instance, as in the case of an atom, only certain energies are allowed to a nucleus. Quantum jumps occur between the energy levels, with the emission or absorption of photons. In the case of nuclei, these are gamma-ray photons, which have energies around 1 MeV (remember 1 MeV = 10^6 eV), i.e. several hundred thousand times greater than those of a visible photon (with energies of just a few eV). Another similarity is that, as in the case of electrons in atoms, the positions and velocities of nucleons in nuclei are subject to indeterminacy. The distinctive feature of nuclei is that the nucleons are confined in a volume that is less than 10^{-14} m in diameter – more than 10^4 times smaller than the diameter of an atom.

However, the possibilities for changes in nuclei are far more varied than those open to electrons in atoms. The big difference is that nuclei can also *transform* from one element to another in a variety of nuclear decays and nuclear reactions. In Book 6 (Section 2.5), you saw how rocks and fossils may be dated by measuring the relative proportions of radioactive parent and daughter nuclei in certain minerals. This chapter examines how these radioactive transformations occur.

Nuclear decays are spontaneous changes that occur in radioactive isotopes, causing the individual nuclei to transform from one type to another. Photons and particles may both be emitted as a result of these decays, carrying away energy in the process. Nuclear decay processes have been known about since the end of the 19th century, from the work of scientists such as Henri Becquerel, Ernest Rutherford and Marie and Pierre Curie. This chapter discusses the characteristic features of all radioactive decays, before focusing in turn on the three types of radioactive decay process that these eminent scientists discovered: alpha-decay (α-decay), beta-decay (β-decay) and gamma-decay (γ-decay).

Nuclear reactions refer more generally to processes in which two or more nuclei are involved, again resulting in the creation of different nuclei from those originally present. In particular, the latter part of this chapter focuses on **nuclear fission** (in which nuclei are split apart) and **nuclear fusion** (in which nuclei are joined together), releasing energy in each case.

6.1 Mass and energy revisited

As you will see later, in all nuclear processes certain conservation rules are obeyed, which are summarised briefly here:

- *electric charge* is always conserved: the net charge of the products of a nuclear process is the same as the net charge of the original nucleus or nuclei
- *mass number* is conserved: the total number of nucleons in the products is the same as that in the original nucleus or nuclei
- as in all physical processes, *energy* is conserved.

The one complication here though, is that since the energies involved in nuclear processes are so large, the relationship between energy and mass, $E = mc^2$, has to be taken into account. Remember from Chapter 2 that it is often convenient to express masses in units of 'energy/c^2', such as MeV/c^2.

The question arises: *why* do certain nuclei transform from one type to another? As you might guess, it is all to do with energy (or equivalently, mass). The first thing to note is that the mass of *any* nucleus is *less than* the total masses of the individual protons and neutrons of which it is composed. For instance, a helium-4 nucleus has a mass of 3727.4 MeV/c^2, whereas the two protons and two neutrons of which it is composed, individually, have a total mass of:

$$(2 \times 938.28)\ \text{MeV}/c^2 + (2 \times 939.57)\ \text{MeV}/c^2 = 3755.7\ \text{MeV}/c^2.$$

The mass difference of:

$$(3755.7 - 3727.4)\ \text{MeV}/c^2 = 28.3\ \text{MeV}/c^2$$

is referred to as the **mass defect** of the helium-4 nucleus. It is the amount by which the nucleus is less massive than its constituent nucleons. Similarly, the **binding energy** of the helium-4 nucleus is −28.3 MeV (note the minus sign!). In forming the helium-4 nucleus from its constituent nucleons, 28.3 MeV of energy is released, as a result of the mass decrease. Conversely, in order to split the helium-4 nucleus apart, 28.3 MeV of energy must be supplied to create the extra mass of the constituent nucleons. This is a similar concept to the ionisation energy that must be supplied to an atom in order to separate an electron from the rest of the atom.

Different nuclei have different mass defects and therefore different amounts of binding energy. Those with the *highest* mass defect have the *lowest* (i.e. most negative) binding energy – both characterised by the same numerical value, as indicated above for helium-4. As you might guess, the binding energy of heavy nuclei is generally lower (i.e. more negative) than the binding energy of light nuclei, simply because they have more nucleons to separate. However, a useful quantity is the **binding energy per nucleon**, calculated simply as the binding energy of the nucleus divided by the number of nucleons it contains. This quantity indicates how strongly each nucleon is bound to the nucleus. For helium-4, which is composed of four nucleons, the binding energy per nucleon is:

$$\frac{-28.3\ \text{MeV}}{4} = -7.1\ \text{MeV}$$

The isotope with the lowest (i.e. most negative) binding energy per nucleon, and therefore the most stable to decay, is iron-56, which has a binding energy per nucleon of −8.8 MeV. The most massive stable isotope is bismuth-209; it has the lowest binding energy of any stable nuclide, with a value of −1640 MeV. Figure 6.1 shows a graph of the binding energy per nucleon versus mass number for a range of stable nuclei. As you can see, iron-56 sits at the lowest point of this graph.

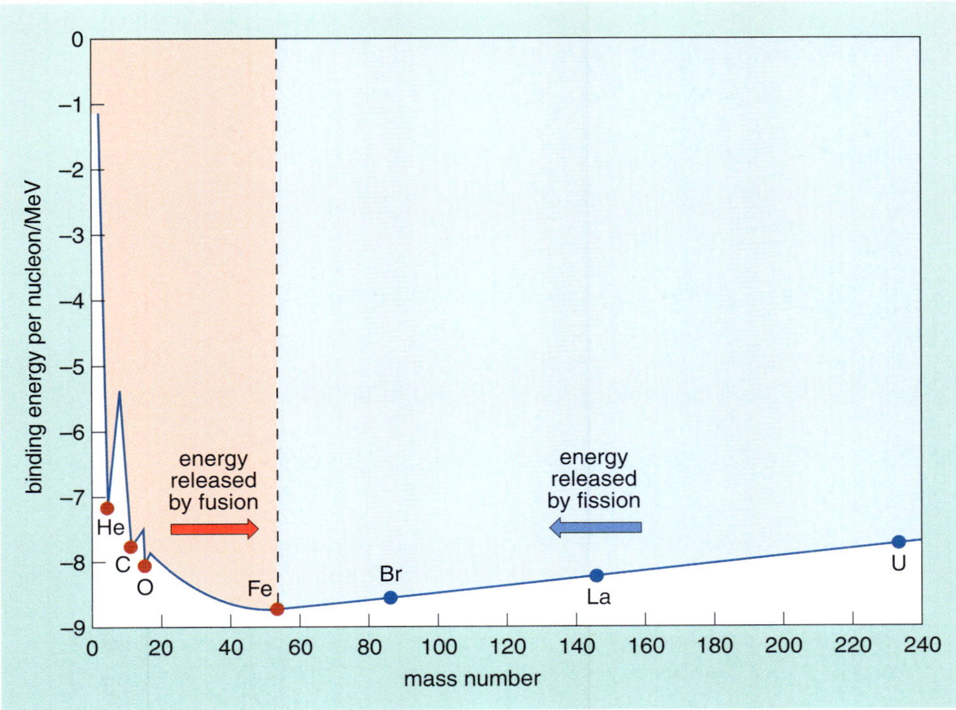

Figure 6.1 The binding energy per nucleon for a range of stable nuclei.

■ (a) What is the binding energy of iron-56? (b) What is the binding energy per nucleon of bismuth-209?

□ (a) Since iron-56 contains 56 nucleons (i.e. the total number of protons and neutrons), its binding energy is $56 \times (-8.8\ \text{MeV}) = -493\ \text{MeV}$. (b) Since bismuth-209 contains 209 nucleons, its binding energy per nucleon is:

$$\frac{-1640\ \text{MeV}}{209} = -7.8\ \text{MeV}$$

6.2 Alpha-decay

An alpha-particle (α-particle) is the same as a helium nucleus, ${}^{4}_{2}\text{He}$, with mass number $A = 4$ and atomic number $Z = 2$. It consists of two protons and two neutrons. As noted above, it is a very tightly bound arrangement; this ground state for two protons and two neutrons has an energy that is 28.3 MeV lower than the energy of the four free nucleons of which it is composed.

In some cases, it is energetically favourable for a nucleus of mass number A and atomic number Z to move to a lower binding energy by emitting an alpha-particle and thereby producing a new nucleus, with mass number $A - 4$ and atomic number $Z - 2$. A case in point is the unstable isotope of uranium, ${}^{234}_{92}\mathrm{U}$.

■ What does this notation indicate for the atomic number and mass number of a uranium nucleus? How many protons and how many neutrons does ${}^{234}_{92}\mathrm{U}$ contain?

□ The notation indicates that the atomic number $Z = 92$ and the mass number $A = 234$. This isotope of uranium has 92 protons and $(234 - 92) = 142$ neutrons.

The uranium-234 nucleus undergoes alpha-decay to produce an isotope of thorium, with 90 protons and 140 neutrons:

$$ {}^{234}_{92}\mathrm{U} \longrightarrow {}^{230}_{90}\mathrm{Th} + {}^{4}_{2}\mathrm{He} $$

Notice that the values for A and Z are the same on the left-hand side of this equation (234 and 92 respectively) as the totals for A and Z on the right hand side ($230 + 4$ and $90 + 2$ respectively), which bears out the assertions earlier that electric charge and mass number are both conserved in nuclear decays.

■ Alpha-decay processes such as this happen a further four times. What isotope of lead results?

□ Since there are four more alpha-decays, the total decrease in mass number is $(5 \times 4) = 20$, and the total decrease in atomic number is $(5 \times 2) = 10$. So the resultant nucleus has $A = 234 - 20 = 214$ and $Z = 92 - 10 = 82$. The element with atomic number 82 is lead, as noted in the question, so the resulting nucleus is the lead isotope ${}^{214}_{82}\mathrm{Pb}$.

Both electric charge and mass number are conserved in an alpha-decay process since there is the same number of protons and neutrons in the products as in the original nucleus. But what about energy conservation? In the example above, the binding energy of the uranium-234 nucleus is –1778.6 MeV, whilst the binding energies of the thorium-230 and helium-4 nuclei are –1755.1 MeV and –28.3 MeV respectively. The total binding energy of the products is therefore –1783.4 MeV. The reaction involves moving to a configuration in which the binding energy has *decreased* by (–1778.6 MeV) – (–1783.4 MeV) = 4.8 MeV. So in this case, the alpha-decay of ${}^{234}_{92}\mathrm{U}$ liberates 4.8 MeV of energy. This takes the form of kinetic energy and is carried away (almost exclusively) by the alpha-particle. So the decay can be written as:

$$ {}^{234}_{92}\mathrm{U} \longrightarrow {}^{230}_{90}\mathrm{Th} + {}^{4}_{2}\mathrm{He} + 4.8\ \mathrm{MeV} $$

It can therefore also be said that the combined mass of the thorium-230 and helium-4 nuclei is 4.8 MeV/c^2 less than the mass of the original uranium-234 nucleus.

Question 6.1

Using the fact that the binding energy of an alpha particle (i.e. a helium-4 nucleus) is −28.3 MeV, answer the following.

(a) The naturally occurring radium isotope (binding energy = −1731.6 MeV) undergoes alpha-decay to a radon isotope (binding energy = −1708.2 MeV). How much energy is liberated by the decay? (Give your answer in MeV.)

(b) The naturally occurring isotope of protactinium (binding energy = −1759.9 MeV) undergoes alpha-decay to produce a nucleus of actinium (binding energy = −1736.7 MeV). By how much is the combined mass of the actinium nucleus and the alpha particle less than the mass of the original protactinium nucleus? (Give your answer in units of MeV/c^2.)

6.3 Beta-decay

This section looks at the second mode of nuclear decay, namely beta-decay. In fact, there are three related processes, each of which is a type of beta-decay. The first one you will consider is **beta-minus (β^-) decay** which involves the emission of an *electron* from the nucleus of an atom. Perhaps the first question to address is: where did that electron come from? Was it there in the first place? The answer is no; it is quite impossible for nuclei to 'contain' electrons. The electron is *created*, by the decay, just as a photon is created when an atom makes a transition from a higher energy level to a lower energy level.

In order to see how the electron comes into existence, a specific case will be considered, namely the beta-minus decay of the lead isotope $^{214}_{82}\text{Pb}$. The electric charge of the lead nucleus is $+82e$ since it contains 82 protons, and the electron emitted by the nucleus has a charge of $-e$. Since charge is conserved in all nuclear decay processes, the charge of the nucleus *before* the decay must be equal to the charge of the nucleus *after* the decay plus the charge of the electron that is emitted. So, $+82e$ = (charge of resultant nucleus) + $(-e)$, and the charge of the resultant nucleus must be $+83e$.

■ How many protons does the resultant nucleus contain? Using the Periodic Table from Book 4 (Figure 5.3), what element does it represent?

□ The resultant nucleus must contain 83 protons. From the Periodic Table, it is therefore a nucleus of the element bismuth ($Z = 83$).

So the resultant bismuth nucleus has one more proton than the original lead nucleus. But where has that proton come from? To answer that you need to consider the mass number of the nucleus. Since mass number is also conserved in nuclear decays, the total number of nucleons (i.e. protons plus neutrons) in the resultant bismuth nucleus must be the same as in the original lead nucleus.

■ How many nucleons does the resultant bismuth nucleus contain, and what is the symbol for this isotope?

□ It contains 214 nucleons – the same as the original lead nucleus. The resultant nucleus is therefore the bismuth isotope $^{214}_{83}\text{Bi}$.

Since the bismuth nucleus contains one more proton than the lead nucleus, but the same number of nucleons in total, it must contain one less neutron than the original lead nucleus. What has happened is that one of the neutrons in the lead nucleus has transformed into a proton, with the emission of an electron. Transformations between neutrons and protons lie at the heart of all beta-decay processes.

For reasons that will become clearer in Chapter 12, another particle is created in the beta-decay process too. It is called the **electron antineutrino** and it has zero electric charge. An electron and an electron antineutrino are always created in a beta-minus decay, the minus sign indicating that the electron is negatively charged. The overall decay process for this lead isotope is therefore:

$$^{214}_{82}\text{Pb} \longrightarrow {}^{214}_{83}\text{Bi} + \text{e}^- + \overline{\nu}_\text{e}$$

The rather clumsy symbol $\overline{\nu}_\text{e}$ represents an electron antineutrino, where ν is the Greek letter *nu* (pronounced 'new'). The subscript 'e' indicates that it is associated with an electron, and the bar over the top of the letter indicates that it is an antimatter particle. As in any nuclear decay, the mass of the products is less than the mass of the original nucleus, and the difference in mass is liberated as kinetic energy of the products.

Question 6.2

A nucleus of the unstable nitrogen isotope $^{16}_{7}\text{N}$ undergoes beta-minus decay. Write down an expression for this nuclear decay, indicating what nucleus is formed as a result. (Refer to the Periodic Table in Book 4 if necessary.)

The process described above is only part of the story as far as beta-decay is concerned. There is a very closely related process, called **beta-plus (β^+) decay**, in which a positron (i.e. an antielectron), is created, along with an **electron neutrino**, which has zero charge. In this process, a proton in the original nucleus transforms into a neutron, so *decreasing* the atomic number by one. A nucleus that undergoes beta-plus decay is the unstable oxygen isotope $^{14}_{8}\text{O}$ which transforms into a stable nitrogen isotope $^{14}_{7}\text{N}$. The decay in this case can be represented as:

$$^{14}_{8}\text{O} \longrightarrow {}^{14}_{7}\text{N} + \text{e}^+ + \nu_\text{e}$$

Here, the symbol e^+ is used to represent the positron and ν_e is the electron neutrino.

Question 6.3

A nucleus of the unstable phosphorus isotope $^{30}_{15}\text{P}$ undergoes beta-plus decay. Write down an expression for this decay process, indicating what nucleus is formed as a result. (Refer to the Periodic Table in Book 4 if necessary.)

The final process in the suite of beta-decays is that of **electron capture**. As suggested by its name, in this process a nucleus captures an electron, usually from the inner regions of the electron cloud surrounding it. A proton in the nucleus interacts with the captured electron, forming a neutron and emitting an electron neutrino. As in beta-plus decay, a proton in the original nucleus transforms into a neutron, so *decreasing* the atomic number by one. A nucleus that undergoes electron capture is the unstable aluminium isotope $^{26}_{13}\text{Al}$, which transforms into a stable magnesium isotope $^{26}_{12}\text{Mg}$. The decay in this case can be represented as:

$$^{26}_{13}\text{Al} + \text{e}^- \longrightarrow {}^{26}_{12}\text{Mg} + \nu_\text{e}$$

Question 6.4

Write a few sentences to compare and contrast the processes of beta-minus decay and beta-plus decay. How does the process of electron capture differ?

6.4 Gamma-decay

The final type of nuclear decay considered here is gamma-decay. In contrast to the two processes of alpha- and beta-decay, this involves no change in the numbers of neutrons and protons. Gamma-decay occurs when a nucleus finds itself in an excited state. A quantum jump down to a lower-energy quantum state, with the same number of neutrons and protons, is accompanied by the emission of a photon, as with transitions in atoms. This time, however, the photon energy is around a million times larger – it is a gamma-ray (γ) photon. Such excited states of nuclei may be created as a result of alpha-decay or beta-decay processes, or by the collisions of nuclei at high kinetic energies.

■ The unstable isotope of caesium, $^{137}_{55}\text{Cs}$, undergoes beta-minus decay to produce an excited state of the barium isotope, $^{137}_{56}\text{Ba}$. The barium nucleus then decays to its ground state with the emission of a gamma-ray photon of energy 662 keV. What are the atomic number and mass number of the barium nucleus *after* the gamma-decay?

□ In the process of gamma-decay, the number of protons and neutrons in the nucleus remains unchanged. So the atomic number and mass number of the barium nucleus after the gamma-decay are the same as they were before, namely 56 and 137 respectively.

6.5 Nuclear fission

As noted in Book 3, Section 8.2, nuclear fission is the name given to the process in which a relatively massive nucleus splits apart into two less massive nuclei, of roughly equal size. In some types of nuclei, this process may occur spontaneously, whilst in others it may be induced by bombarding a nucleus with other particles. An example will serve to illustrate the principles involved.

If a nucleus of the isotope uranium-235 absorbs a slow-moving neutron, it will temporarily form a nucleus of uranium-236. This nucleus is unstable and so will rapidly split apart to form two smaller nuclei, such as krypton-92 and barium-141, plus a few ‘left-over’ neutrons. The particular nuclei, and specific number of neutrons, formed by each such interaction are random, but these fission products are usually themselves radioactive, and subsequently decay via a range of beta- and gamma-decays. In the case stated above, the reaction may be written as:

$$^{235}_{92}\text{U} + ^{1}_{0}\text{n} \longrightarrow ^{236}_{92}\text{U} \longrightarrow ^{92}_{36}\text{Kr} + ^{141}_{56}\text{Ba} + 3^{1}_{0}\text{n}$$

Note that here, *three* neutrons are ejected from the fission process, and a neutron can be written as $^{1}_{0}\text{n}$ to emphasise that it has a charge of zero, but a mass number of one.

■ Verify that electric charge and mass number are conserved in this reaction.

□ The electric charge of the original nucleus is $+92e$, whereas that of the product nuclei is $+36e + 56e = +92e$, as required (neutrons have no charge). The mass number of the reactants is $235 + 1 = 236$, whereas that of the products is $92 + 141 + 3 = 236$, as required.

Now look at the binding energies of the participants in this reaction. The binding energy of uranium-235 is −1783.9 MeV, whilst the binding energies of the krypton-92 and barium-141 nuclei are −783.2 MeV and −1174.0 MeV respectively. The total binding energy of the product nuclei is therefore −1957.2 MeV, and this is significantly lower than that of the original nucleus. The difference in energy, 173.3 MeV, is released principally as the kinetic energy of the free neutrons. If these free neutrons are slowed down, they are able to interact with any further uranium-235 nuclei present in the sample, and so initiate a self-sustaining **chain reaction**.

Such reactions have been employed both in nuclear power stations, where the energy released is used to heat and vaporise water that turns turbines to generate electricity, and in nuclear weapons, where the energy released is put to more destructive use. But such reactions also occur naturally in the Earth. There is a site at Oklo in West Africa (Figure 6.2) where the concentrations of various

Figure 6.2 The site of the naturally occurring nuclear reactor at Oklo, in Gabon, West Africa.

naturally occurring uranium isotopes indicate that a self-sustaining chain reaction occurred here about 1.7 billion years ago. In this natural reactor, it is believed that water filtering through crevices in the rock played a key role. Just as in a nuclear power station, the water slowed down the neutrons that were emitted during the fission process so that they could interact with other uranium nuclei. Without the water to slow them down, the neutrons would move so fast that they would just bounce off and not produce a chain reaction. When the rate of heat production from the reactions became so high that the water vaporised, the neutrons were not slowed down and the reactions stopped until the water cooled again. At this point the process could begin again. The chain reactions stopped completely when the abundance of uranium-235 nuclei became too low to keep the reactions going.

Question 6.5

Suppose that a particular uranium-235 nucleus ($Z = 92$), when struck by a slow-moving neutron, produces a nucleus of strontium-99 ($Z = 38$) and one other nucleus, and two free neutrons. What are the mass number and atomic number of the other nucleus produced?

6.6 Nuclear fusion

As noted in Book 3, Section 8.2, nuclear fusion is the name given to the process by which two (or more) low-mass nuclei join together to form a heavier nucleus. In general, these reactions only occur at extremely high temperatures, because the positively charged nuclei need to be 'forced' close enough together in order for the fusion to happen. At low temperatures, the electrical repulsion between them keeps them far enough apart not to interact. Like nuclear fission, nuclear fusion too is a natural process that occurs throughout the Universe. In fact it is the process that powers all stars, including the Sun, and is largely responsible for creating the range of elements that make up the world around us. You will look at the consequences of fusion reactions in the Universe more closely in Chapter 15, but for now you will consider a particular set of reactions – those occurring in the Sun and other low-mass stars – and see how they can be used to illustrate the principles involved.

Energy is generated in the core of the Sun by a set of nuclear fusion reactions known as the **proton–proton chain** (or PP chain for short). The chain begins with two protons (hydrogen nuclei) fusing together to make a nucleus of deuterium (heavy hydrogen).

$$ {}^{1}_{1}\mathrm{H} + {}^{1}_{1}\mathrm{H} \longrightarrow {}^{2}_{1}\mathrm{H} + \mathrm{e}^{+} + \nu_{\mathrm{e}} \qquad \text{(step 1)} $$

The net effect of this reaction is that one of the original protons has transformed into a neutron with the emission of a positron and an electron neutrino. This is therefore similar to the process of beta-plus decay which you met in Section 6.3. Because of the conversion of a proton into a neutron, it cannot just be assumed that the binding energy of the deuterium nucleus (–2.2 MeV) is released here. Instead you will note that the reactants have a mass-energy of (2 × 938.3 MeV) = 1876.6 MeV, and the products have a mass-energy of 1875.7 MeV (for the deuterium) + 0.5 MeV (for the positron) plus a negligible amount for the electron neutrino, i.e. a total of 1876.2 MeV. So, the amount of energy liberated is (1876.6 MeV − 1876.2 MeV) = 0.4 MeV. In addition, the positron will immediately annihilate with an electron to give a further 2 × 0.5 MeV = 1.0 MeV in the form of photons. So the total energy release from this reaction is (0.4 MeV + 1.0 MeV) = 1.4 MeV.

The next step in the chain is that the deuterium nucleus quickly captures another proton (hydrogen nucleus), to make a nucleus of helium-3:

$$^{2}_{1}\mathrm{H} + {}^{1}_{1}\mathrm{H} \longrightarrow {}^{3}_{2}\mathrm{He} \qquad \text{(step 2)}$$

Here, the binding energy of the reactants is −2.2 MeV (deuterium) + 0 MeV (hydrogen), and that of the product is −7.7 MeV (helium-3). So a further 5.5 MeV of energy is released in this step.

The final step is that two of these helium-3 nuclei fuse together to form a nucleus of helium-4, releasing two protons (hydrogen nuclei) back into the mix:

$$^{3}_{2}\mathrm{He} + {}^{3}_{2}\mathrm{He} \longrightarrow {}^{4}_{2}\mathrm{He} + {}^{1}_{1}\mathrm{H} + {}^{1}_{1}\mathrm{H} \qquad \text{(step 3)}$$

■ Using information from earlier, what is the energy released by this final step in the reaction?

□ The binding energy of helium-3 is −7.7 MeV and the binding energy of helium-4 is −28.3 MeV. So, the binding energy of the reactants is (−7.7 MeV) + (−7.7 MeV) = −15.4 MeV, whilst that of the products is (−28.3 MeV) + 0 MeV. Therefore an energy of (−15.4 MeV) − (−28.3 MeV) = 12.9 MeV is released.

Since two instances of each of the first and second steps in the chain are needed for each instance of the third step in the chain (in order to make the two helium-3 nuclei required), the complete reaction is that six protons have been combined to make a helium-4 nucleus plus two protons, two positrons and two electron neutrinos. The two positrons in turn will annihilate with two electrons, producing photons:

$$6\,^{1}_{1}\mathrm{H} \longrightarrow {}^{4}_{2}\mathrm{He} + 2\,^{1}_{1}\mathrm{H} + 2\mathrm{e}^{+} + 2\nu_{\mathrm{e}}$$

and

$$2\mathrm{e}^{+} + 2\mathrm{e}^{-} \longrightarrow \textit{photons}$$

The net effect is therefore that four protons and two electrons have been combined to make a single helium-4 nucleus and two electron neutrinos, plus some photons:

$$4\,^{1}_{1}\mathrm{H} + 2\mathrm{e}^{-} \longrightarrow {}^{4}_{2}\mathrm{He} + 2\nu_{\mathrm{e}} + \textit{photons}$$

The complete proton–proton chain is illustrated in Figure 6.3. The total energy release is therefore (1.4 MeV × 2) from step 1, plus (5.5 MeV × 2) from step 2, plus 12.9 MeV from step 3. Adding these together gives 26.7 MeV. Notice that this is slightly less than the binding energy of the helium-4 nucleus (28.3 MeV), as it has been created from four protons rather than two protons and two neutrons.

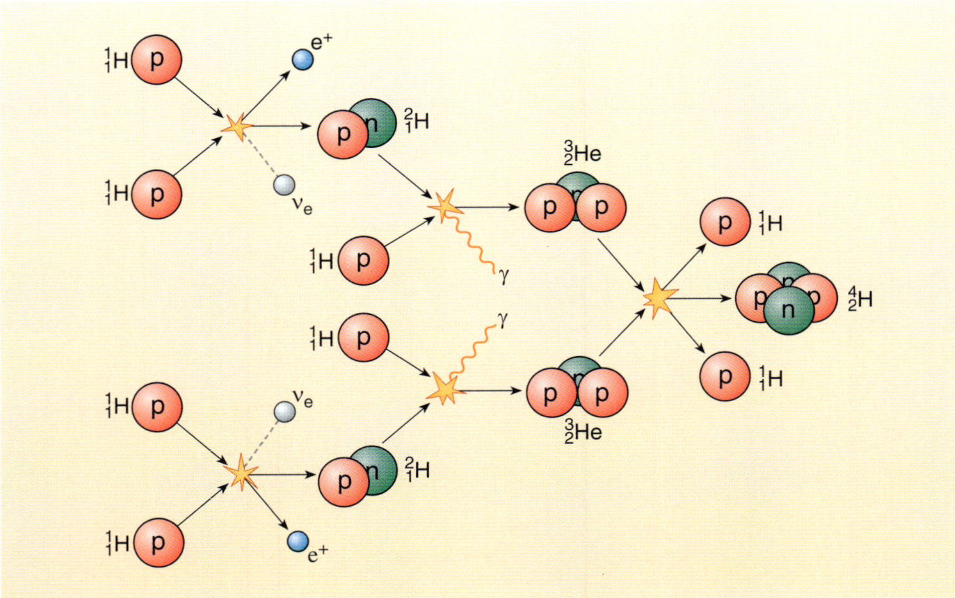

Figure 6.3 The reactions comprising the proton–proton chain.

In general, fusion reactions to build more massive nuclei from lighter ones will release energy all the way up to the nucleus iron-56, which, as was noted earlier, has the lowest (most negative) binding energy per nucleon. As you will see in Chapter 15, this proves to be crucial in determining the life cycles of stars.

Question 6.6

When stars run out of hydrogen in their cores, they may begin a new type of fusion reaction called the **triple-alpha process** in which three helium-4 nuclei fuse together to make a single nucleus of carbon-12.

(a) Write an equation for this fusion process using the usual symbols for the nuclei involved.

(b) Given that the binding energy of a carbon-12 nucleus is –92.2 MeV, how much energy is liberated by this process?

6.7 Summary of Chapter 6

In all nuclear decays, electric charge, mass number and energy are conserved. Because of the high energies involved, the conversion between mass and energy must be taken into account. It is often useful to express the masses of particles in units of MeV/c^2 to emphasise the energy changes that occur.

The mass of any nucleus is less than the total mass of the individual nucleons of which it is composed. The difference in mass, known as the mass defect, may also be expressed as the binding energy of the nucleus. A helium-4 nucleus has a mass defect of 28.3 MeV/c^2 or a binding energy of –28.3 MeV. Iron-56 is the most stable nucleus, with a binding energy per nucleon of –8.8 MeV.

There are a total of five types of nuclear decay or transition:

- alpha-decay in which helium nuclei are emitted
- beta-minus decay in which a neutron converts into a proton with the emission of an electron and an electron antineutrino
- beta-plus decay in which a proton converts into a neutron with the emission of a positron and an electron neutrino
- electron capture in which a proton captures an electron so turning into a neutron and emitting an electron neutrino
- gamma-decay in which a nucleus in an excited state makes a transition to a lower energy level, accompanied by the emission of a gamma-ray photon.

Nuclear fission involves massive nuclei splitting apart to form two or more less massive nuclei, with the release of energy. Nuclear fusion involves two or more low-mass nuclei joining together to form a more massive nucleus, with the release of energy.

Chapter 7
Fundamental particles

> It was here that the thaum, hitherto believed to be the smallest possible particle of magic, was successfully demonstrated to be made up of *resons* (Lit.: 'Thing-ies') or reality fragments. Currently research indicates that each reson is itself made up of a combination of at least five 'flavours', known as 'up', 'down', 'sideways', 'sex appeal' and 'peppermint'.
>
> Terry Pratchett, British author (1948–) in *Lords and Ladies*

Half a century ago, the account of subatomic structure would have ended with nucleons. But now a third layer of structure is known: it is believed that protons and neutrons are composed of structureless particles known as quarks. This chapter will be looking at an area that is at the forefront of scientific research today. The aim here is nothing less than an understanding of the fundamental constituents from which the Universe is built.

So far in this book, you have met four types of subatomic particle: electrons, electron neutrinos, protons and neutrons, along with their antiparticles in some cases. Over the last 50 years or so, experiments have revealed that other subatomic particles exist, and that those mentioned above are merely the common representatives of two distinct classes of object, known as leptons and hadrons. These are now considered in turn, in order to complete the final leg of your journey to take the world apart.

7.1 Leptons

Electrons (e^-) and electron neutrinos (ν_e), together with their antiparticles, are believed to be **fundamental particles**. By fundamental, is meant that there is no evidence that they are composed of smaller or simpler constituents. Furthermore, two more particles, with the same charge as the electron, only rather heavier, were discovered in 1936 and 1975. The first is known as the muon (represented by μ^-; the Greek letter *mu*, rhymes with 'cue') which is about 200 times heavier than the electron. The second is called the tauon (represented by τ^-; the Greek letter *tau*, rhymes with 'cow') which is about 3500 times heavier than the electron. The superscript minus signs on the electron, muon and tauon indicate that these particles have negative electric charge. Like the electron, the muon and tauon each has an associated neutrino: the muon neutrino (ν_μ) and the tauon neutrino (ν_τ).

These six fundamental particles are collectively referred to as **leptons** – they are listed in Table 7.1 along with their electric charge. (The word lepton comes from the Greek *leptos*, meaning 'thin' or 'lightweight'.) The six different types are often rather whimsically referred to as different **flavours** of lepton, and the three pairs of particles are often referred to as three **generations** of leptons.

Table 7.1 Three generations of lepton.

	1st generation	2nd generation	3rd generation
leptons with charge $-e$	e^-	μ^-	τ^-
leptons with charge 0	ν_e	ν_μ	ν_τ

To each lepton there corresponds an **antilepton** with opposite charge but with the same mass. These are denoted by the symbols e^+, μ^+ and τ^+ for the charged leptons and $\overline{\nu}_e$, $\overline{\nu}_\mu$ and $\overline{\nu}_\tau$ for the neutral leptons.

Until very recently, it was not known whether the three types of neutrino possess mass, like the charged leptons, or whether they are massless, like photons. However, in 2005 results from studying the neutrinos emitted by the Sun (Figure 7.1) showed that they undergo 'flavour oscillations' – some of the electron neutrinos produced by the proton–proton chain in the core of the Sun *change* into muon neutrinos or tauon neutrinos as they travel through space before reaching the Earth. This is the solution to the long-standing problem of why fewer electron neutrinos are detected from the Sun than are believed to be produced in its core. The implication of this is that neutrinos do indeed possess mass, as massless neutrinos would not be able to change flavour, according to current theories of how these particles behave. Measurements show that the *total* mass of the electron neutrino, muon neutrino and tauon neutrino must be less than 0.3 eV/c^2 and that the most massive of the three neutrinos must have a mass of at least 0.05 eV/c^2.

Figure 7.1 The Sudbury Neutrino Observatory in Canada consists of a huge underground tank of 'heavy water' (i.e. water containing deuterium rather than normal hydrogen) surrounded by sensitive light detectors. When neutrinos from the Sun interact with the deuterium, tiny flashes of light are recorded that enable the reactions to be studied.

7.2 Quarks

The other subatomic particles that you have met so far – protons and neutrons – are examples of **hadrons**. (The word hadron comes from the Greek *hadros*, meaning 'strong' or 'robust'.) Although the only hadrons existing around us in the everyday world are protons and neutrons, many more types of hadron can be created in high-energy collisions of nucleons. Such reactions are common in the upper atmosphere, where high-energy protons from outer space (known as cosmic-ray protons) collide with nuclei of nitrogen and oxygen, smashing them apart and creating new hadrons. Since the 1960s, such reactions have been closely studied under controlled conditions, in high-energy physics laboratories such as that shown earlier in Figure 1.3, where protons and electrons are accelerated to high kinetic energies using very high voltages.

Although many dozens of different types of hadron may be created in this way, all of the new ones are unstable and they rapidly decay into other, long-lived particles, such as leptons, protons and neutrons. Fortunately, it will not be necessary for you to dwell on (let alone remember) the names and properties of all the types of hadron, because there is a straightforward description for building them from particles that *are* believed to be fundamental, namely from **quarks** and **antiquarks**.

There are six flavours of quark, labelled by the letters u, d, c, s, t and b, which stand for up, down, charm, strange, top and bottom. (The last two are sometimes referred to as 'truth' and 'beauty' instead of 'top' and 'bottom'. Despite the quote by Terry Pratchett at the beginning of this chapter, there are no quarks called 'sex appeal' or 'peppermint', although the strange quark was for a time known as 'sideways'!) The charges of the quarks, in units of the proton charge, e, are listed in Table 7.2. Like the leptons, the six quarks are often grouped into three generations, on the basis of their mass, with the first generation being the least massive.

Table 7.2 Three generations of quark.

	1st generation	2nd generation	3rd generation
quarks with charge $+\frac{2}{3}e$	u	c	t
quarks with charge $-\frac{1}{3}e$	d	s	b

To each quark, there corresponds an antiquark, with the opposite charge and the same mass. These are denoted by $\overline{\mathrm{u}}$, $\overline{\mathrm{d}}$, $\overline{\mathrm{c}}$, $\overline{\mathrm{s}}$, $\overline{\mathrm{t}}$ and $\overline{\mathrm{b}}$.

The **up quarks** and **down quarks** are the constituents of protons and neutrons, and along with their antiquark counterparts are the least massive of all the quarks. The charm and strange quarks and antiquarks are more massive than the up and down quarks, and the top and bottom quarks and antiquarks are yet more massive still. The large masses of these second- and third-generation quarks are the reason why powerful particle accelerators are required to produce them. In order to create this amount of mass, a large amount of kinetic energy must be supplied in accordance with Equation 2.1. In fact, hadrons containing top quarks were first detected in the period 1994–5. They have masses nearly 200 times that of the proton.

Quarks and antiquarks have *never* been observed in isolation. They only occur bound together inside hadrons. In fact there are three confirmed recipes for building hadrons from quarks, and these are shown in Figure 7.2.

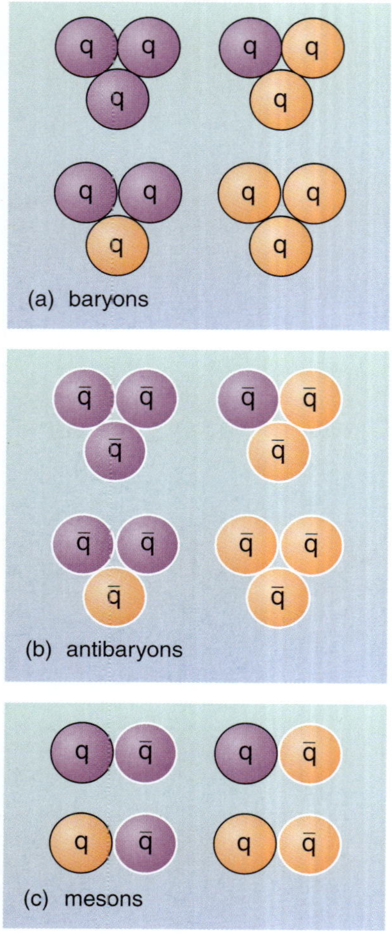

Figure 7.2 The three recipes for building hadrons from quarks. Quarks and antiquarks with a charge of $\pm\frac{2}{3}e$ are shown in purple, those with a charge of $\pm\frac{1}{3}e$ are shown in orange. The symbol q represents a quark, and $\bar{q}$ represents an antiquark. (a) Possible combinations making a baryon; (b) possible combinations making an antibaryon; (c) possible combinations making a meson.

A hadron can consist of:

- three quarks (in which case it is called a **baryon**) or
- three antiquarks (in which case it is called an **antibaryon**) or
- one quark and one antiquark (in which case it is called a **meson**).

The net electric charge of a hadron is simply the sum of the electric charges of the quarks or antiquarks of which it is composed. Notice that the net charge of a hadron is therefore *always* a whole number, despite the fact that the quarks themselves have non-whole number electric charge.

As a specific example of the hadron-building recipe, the proton is a baryon, so it is composed of three quarks, and as mentioned above, it is composed of up and down quarks only. Now, the proton has charge $+e$ and the only way that three up or down quarks can be combined to make this net charge is by combining two up quarks with a down quark. So the quark content of a proton is (uud), giving a charge $\frac{2}{3}e+\frac{2}{3}e-\frac{1}{3}e=\frac{3}{3}e=+e$.

■ Following the pattern above, determine the quark content of a neutron.

□ Since the neutron has charge 0 and is composed of three up or down quarks, its quark content must be (udd), giving a charge $\frac{2}{3}e-\frac{1}{3}e-\frac{1}{3}e=0$.

In recent years, this standard picture of baryons, antibaryons and mesons has been challenged by new experimental results. In 2003, experiments at particle accelerators in Japan and the USA reported the detection of a new type of short-lived particle referred to as an 'exotic baryon' or 'pentaquark' (literally '5 quarks') composed of four quarks plus one antiquark. Then, in 2007, a Japanese team reported evidence for a new type of meson comprising two quarks and two antiquarks. Each of these new particles decays very rapidly into other, more familiar hadrons, making them very difficult to study. At the time of writing (late 2007), these discoveries are still awaiting confirmation, but they illustrate that particle physics is still a rapidly moving field of research and that scientists may not (yet) have the full picture.

A summary of the nomenclature introduced in this chapter is given in Table 7.3. The other 'particle' that you have met so far, namely the quantum of electromagnetic radiation (i.e. the photon), is rather different in nature to these particles. You will return to classifying it, along with other quanta, in Chapters 10–13.

Table 7.3 Fundamental particles and the composite particles made from them.

<table>
<tr><th colspan="2">Leptons</th><th colspan="2">Quarks</th></tr>
<tr><td rowspan="4">electron

muon

tauon

electric charge $= -e$</td><td rowspan="4">electron neutrino

muon neutrino

tauon neutrino

electric charge $= 0$</td><td>up, charm, top quarks

electric charge $= +\frac{2}{3}e$</td><td>down, strange, bottom quarks

electric charge $= -\frac{1}{3}e$</td></tr>
<tr><th colspan="2">Hadrons</th></tr>
<tr><th>Baryons</th><th>Mesons</th></tr>
<tr><td>composed of three quarks</td><td>composed of quark</td></tr>
</table>

This tally of six leptons and six quarks, each with their own antiparticles, may seem like a huge number of fundamental particles. However, don't let this put you off. Everything around us is made up of merely the first generation of each type, namely electrons, up quarks and down quarks, with electron neutrinos being created in beta-decays. The second generation of leptons (μ^- and ν_μ) and the second generation of quarks (c and s), the third generation of leptons (τ^- and ν_τ) and the third generation of quarks (t and b), have exactly the same properties as their first-generation counterparts except that they are more massive. Quite why nature decided to repeat this invention three times over is not currently understood, but you may be pleased to learn that current theories do not predict the existence of any more generations.

Question 7.1

Antiparticles of the proton and neutron are composed of antiquarks. What are the constituents of (a) an antiproton and (b) an antineutron? In each case calculate the charge of the resulting antiparticle by adding up the charges of the constituent antiquarks.

7.3 High-energy reactions

In particle accelerators, or in the Earth's upper atmosphere, hadrons can smash into each other with large kinetic energies, and create new hadrons from the debris left behind. An illustration of the complex trails left by particles as they react in a particle accelerator is shown in Figure 7.3.

Figure 7.3 Trails left by particles as they interact in a particle accelerator.

As an example of the high-energy reactions that can occur, when a proton with kinetic energy of several hundred MeV collides with a nucleon, new hadrons, called **pions**, can be created. In such processes, some of the kinetic energy of the protons is converted into mass, via the familiar equation $E = mc^2$, and so appears in the form of new particles. The pions that are created come in three varieties: π^+, with positive charge $+e$, π^- with negative charge $-e$, and a neutral π^0 with zero charge. Their masses are each around 140 MeV/c^2, so 140 MeV of kinetic energy is required in order to create each pion. Pions are examples of mesons and so are composed of a quark and an antiquark. In fact, pions are the least massive mesons and (like protons and neutrons) are composed of up and down quarks. To get a positive charge on π^+, the combination $(\mathrm{u}\overline{\mathrm{d}})$ is needed, with charge $+\frac{2}{3}e+\frac{1}{3}e=+e$.

■ What is the quark–antiquark content of the negatively charged pion, π^-?

□ It must have the combination $(\mathrm{d}\overline{\mathrm{u}})$, with charge $-\frac{1}{3}e+(-\frac{2}{3}e)=-e$.

But now there arises an interesting question about π^0: is it $(\mathrm{u}\overline{\mathrm{u}})$ or $(\mathrm{d}\overline{\mathrm{d}})$? Each possibility follows the rules, and each produces zero charge. The general principle of quantum physics seems to be that 'whatever is not strictly forbidden will be found to occur'. Each possibility is allowed; each is found to occur. But that does not mean that are there two types of π^0. It means that if we ask nature, in a suitable experiment, what is the content of π^0, sometimes we get one answer, sometimes the other. You have seen similar things before in the quantum world. For instance, in atomic physics, it is only the act of measurement that forces one of the possible outcomes for the position of an electron. In particle physics, the very notion of what something is made of is subject to the same indeterminacy, until we ask. It is then subject to the same regularity of probabilities, when we do. Satisfyingly, experiments to determine the quark–antiquark content of π^0, by a study of its interactions, give one answer 50% of the time, and the alternative answer the other 50%.

Question 7.2

There is a hadron, formed in the collisions of pions and nucleons, with charge $+2e$. How can a single pion combine with a single nucleon to produce nothing but this new hadron? Is it a baryon, an antibaryon or a meson?

The outcome from a high-energy collision between, say, two protons is subject to quantum indeterminacy in several ways. Imagine that two protons collide with a total kinetic energy of 300 MeV. This is enough energy to make (up to) two pions (remember, they have a mass energy of 140 MeV each) and still have a little energy left over to provide kinetic energy of the products. But what exactly will the products be? In fact there are several possibilities in this case:

$$\mathrm{p}+\mathrm{p} \longrightarrow \mathrm{p}+\mathrm{p}+\pi^0$$

$$\mathrm{p}+\mathrm{p} \longrightarrow \mathrm{p}+\mathrm{n}+\pi^+$$

$$\mathrm{p}+\mathrm{p} \longrightarrow \mathrm{p}+\mathrm{p}+\pi^0+\pi^0$$

$$\mathrm{p}+\mathrm{p} \longrightarrow \mathrm{p}+\mathrm{p}+\pi^++\pi^-$$

$$\mathrm{p}+\mathrm{p} \longrightarrow \mathrm{p}+\mathrm{n}+\pi^++\pi^0$$

$$\mathrm{p}+\mathrm{p} \longrightarrow \mathrm{n}+\mathrm{n}+\pi^++\pi^+$$

Each of these possibilities will occur, and there is no way of predicting which one will be the outcome of a particular reaction. The rules that must be obeyed in these reactions are:

- energy is conserved (as usual)
- electric charge is conserved (as usual)
- the number of quarks minus the number of antiquarks is conserved.

Considering each of these rules in turn for the six reactions above, first look at the conservation of energy. In each case, the mass energy of the two protons on the left-hand side (2×940 MeV) is balanced by the mass energy of the two protons or neutrons on the right hand side (2×940 MeV). So, the only concern is what happens to the 300 MeV of kinetic energy of the reactants. The mass energy of each pion is 140 MeV, and clearly energy will be conserved as long as the kinetic energy of the products is 160 MeV in the first two reactions, or 20 MeV in the last four reactions. (Three pions cannot be created as that would require at least 3×140 MeV $= 420$ MeV of kinetic energy from the reactants.)

The electric charge of the reactants is $+2e$ in each case. The electric charge of the products in the six cases are:

$$+e+e+0=+2e$$

$$+e+0+e=+2e$$

$$+e+e+0+0=+2e$$

$$+e+e+e-e=+2e$$

$$+e+0+e+0=+2e$$

$$0+0+e+e=+2e$$

So, electric charge is conserved in all six cases.

Finally, the reactants are composed of 3 quarks for each proton and no antiquarks, so the number of quarks minus the number of antiquarks is 6 for the reactants. In the products, the number of quarks is 3 for each proton or neutron and 1 for each pion, whilst the number of antiquarks is 1 for each pion. The number of quarks minus the number of antiquarks in the reactants is therefore $(3+3+1-1)=6$ in the first two cases and $(3+3+1-1+1-1)=6$ in the last four cases. Therefore all three rules are indeed obeyed in all six reactions.

Question 7.3

Imagine that two neutrons collide with a combined kinetic energy of 500 MeV. Write down all the possible reactions for producing pions that may occur.

In conclusion, just as electrons, protons and neutrons simplify the elements to their essentials, so a new layer of apparent complexity can be understood by the quark model. However, there is a crucial difference. If an atom is hit hard enough, electrons come out. If a nucleus is hit hard enough, nucleons come out. But if a nucleon is hit hard, with another nucleon, quarks do not come out. Instead the kinetic energy is transformed into the mass of new hadrons.

7.4 Summary of Chapter 7

The electron and the electron neutrino are examples of leptons. There are six types of lepton in total, all of which are believed to be fundamental, structureless particles. There are also six types of quark. The least massive are the up quark and the down quark. Quarks, like leptons, are believed to be fundamental particles. Unlike leptons, they have never been observed in isolation.

Hadrons are composite particles made of quarks and antiquarks. Combinations of three quarks are called baryons, combinations of three antiquarks are called antibaryons, and combinations of a quark and an antiquark are called mesons. Nucleons (protons and neutrons) are merely special cases of baryons. Protons have the quark composition (uud) and neutrons have the quark composition (udd). Pions are examples of mesons, and are also composed of up and down quarks and antiquarks.

High-energy reactions involving hadrons may be interpreted as the rearrangement of quarks accompanied by the creation, or annihilation, of quark–antiquark pairs, in accordance with the usual conservation laws of energy and charge.

Activity 7.1 Reflections on quantum physics

We expect this activity will take you approximately 20 minutes.

You have now completed the first part of the book and have reached the end-point of your quest to examine the Universe on its smallest size scales. This introduction to the world of quantum physics has revealed that all material objects are composed of atoms, which in turn consist of negatively charged electrons surrounding a positively charged nucleus. The nuclei of atoms are made up of protons and neutrons, which in turn are formed from combinations of up and down quarks. So to build everything around us – from the food you had for breakfast and the book or computer screen in front of you, to every person you have ever met, the planet Earth and all the stars and galaxies in the Universe – requires only *three* types of particle: electrons, up quarks and down quarks. Virtually imperceptible particles known as electron neutrinos complete the picture of the material world.

Our understanding of these fundamental building blocks is based on quantum physics. This tells us that we can have no absolute knowledge of the position of an individual particle, or the velocity with which it moves. All we can ever hope to measure is some kind of probability for finding it in a certain place at a certain time. Systems of particles – such as atoms – do have a well-defined energy, however. Their energy is so well defined in fact, that the energy of an atom is only allowed to have certain specific values. Between these allowed energy levels, dramatic quantum jumps occur.

It is these jumps that provide the link with the other important component of the world around us: electromagnetic radiation. When an atom jumps from one energy level to another, such that it loses energy, a photon is emitted whose energy corresponds exactly to the energy difference between the two energy levels. Alternatively, a photon whose energy corresponds exactly to the energy difference between two energy levels of an atom may be absorbed, so raising the energy of the atom and exciting it to the higher energy level.

(a) Go back to Figure 3.2, showing the photograph of the 1927 Solvay International Conference, and for each of the scientists identified in the caption (other than Paul Dirac who will feature later) write a sentence summarising their contribution to our understanding of the behaviour of the Universe on small scales, as outlined in the previous five chapters of this book.

(b) Look around you now and make a list of half a dozen things or processes that rely, ultimately, on the underlying rules of quantum physics operating on atoms and subatomic particles.

Now look at the comments on this activity at the end of this book.

Having reached the limits of human understanding at the smallest length scales, you will now turn to the largest length scales of the Universe and await the surprises in store for you there.

PART II – HOW DOES THE UNIVERSE BEHAVE ON LARGE SCALES?

> I am very interested in the Universe – I am specializing in the Universe and all that surrounds it …
>
> Peter Cook, British comedian (1937–1995), speaking as *E. L. Wisty*

In this second part of the book, some of the key observations in the realm of cosmology are introduced. The structure and contents of our local patch of the Universe are reasonably familiar to many people. As you saw in Book 2, the Earth is a rather small, rocky body and is one of several planets that orbit our local star, the Sun. The Sun itself is a fairly average star, about 5 billion years old and midway through its life. It is just one of a hundred billion or so stars that comprise our galaxy, known as the Milky Way. The Galaxy takes the form of a flattened spiral structure, with the Sun lying in one of the outer spiral arms, about two-thirds of the way from the centre to the edge. Finally, our galaxy is merely one of several dozen that comprise the Local Group of galaxies, and the Local Group is itself part of the much larger Local Supercluster of galaxies that may contain around 100 000 individual galaxies.

In the following two chapters, you will investigate different aspects of the way the Universe appears today. First, you will look at how the Universe is moving on its largest scales as you consider the expanding Universe. Then, you will consider the radiation content of the Universe and examine how the Universe is gradually cooling. So, by the end of this second part of the book, you will have explored answers to the question posed in Chapter 1 – how does the Universe behave on large scales?

Chapter 8 The expanding Universe

> Equipped with our five senses, we explore the Universe around us and call the adventure Science.
>
> Edwin Hubble, American astronomer (1889–1953)

The deduction that the Universe is expanding is based on measurements of two quantities for each of thousands of galaxies: their distance away and the apparent speed with which they are moving. Each of these quantities is determined in a quite straightforward manner by applying laws of physics that are tried and tested here on Earth – but applying them to situations on a much larger scale of both distance and time. In order to make any sense of the observations that will be discussed, it is necessary to assume that the laws of physics that operate in distant parts of the Universe (distant in both time and space) are the *same* as those that operate in laboratories on the Earth, today. In fact, this is only an extreme version of an assumption that underlies the whole of science: it is assumed that the laws of physics were the same in Birmingham yesterday as they will be in Bangalore tomorrow, for instance. If this were not true, then no further progress would be possible. Conversely, the fact that apparently sensible conclusions can be reached by making just this one assumption, tends to indicate that it is not such a bad assumption after all. If such assumptions were to lead to inconsistencies with observations, then we'd have to re-examine the original assumptions and possibly modify the laws of physics as they are currently expressed. This process is the essence of scientific method.

Virtually everything that is known about the properties and behaviour of the Universe at large has been learnt from the light and other electromagnetic radiation emitted by distant stars and galaxies. When light from a galaxy is collected using telescopes, different types of measurements can be made on it. The simplest measurement is to determine how bright the galaxy appears to be, that is, how much light emitted by the galaxy is detected here on Earth. A slightly more complex measurement is to examine the spectrum of light emitted by the galaxy. As you will see in the rest of this chapter, it is measurements of the brightness and the spectrum of a distant galaxy that can lead to determinations of its distance and apparent speed, respectively.

8.1 The distance to galaxies

Astronomical distances are so enormous that metres (or even kilometres) are too small a unit with which to conveniently measure them. For this reason cosmologists usually use a different unit known as the **parsec** (symbol pc) which is equal to about 31 thousand billion kilometres (3.1×10^{13} km). In terms of another unit mentioned earlier, a parsec is equal to about 3.3 **light-years** (symbol ly). In other words, it would take a beam of light just over three years to travel a distance of one parsec. The relationship between these three units is therefore:

$$1.0 \text{ pc} \approx 3.3 \text{ ly} \approx 3.1 \times 10^{13} \text{ km}$$

where the symbol '≈' may be read as 'is approximately equal to'. (All conversions are accurate to two significant figures.)

To give you an idea of the sort of distance scales that are involved in the Universe, the distance from the Earth to the Sun is about 1.5×10^8 km or only 0.000 005 pc, and the Solar System out to the orbit of Pluto has a diameter of about 7.5×10^9 km or 0.000 25 pc. The nearest star to the Sun, Proxima Centauri, is about 4.2 ly or 1.3 pc away. As illustrated in Figure 8.1, our own galaxy, the Milky Way, is about 40 000 pc, or 40 kpc (40 kiloparsecs), in diameter and one of our nearest neighbours, the Andromeda galaxy, is about 660 000 pc or 660 kpc (660 kiloparsecs) distant from us. Clusters of galaxies typically lie at distances of several hundred million parsecs away (one million parsecs is written as 1 Mpc or 1 megaparsec), or even several billion parsecs away (one billion parsecs is written as 1 Gpc or 1 gigaparsec).

■ What is a distance of 1 Gpc expressed in metres?

□ Since 1.0 pc = 3.1×10^{13} km, then 1.0 Gpc = $10^9 \times 3.1 \times 10^{13}$ km = 3.1×10^{22} km. Converting to metres, 1 Gpc = $3.1 \times 10^{22} \times 10^3$ m = 3.1×10^{25} m.

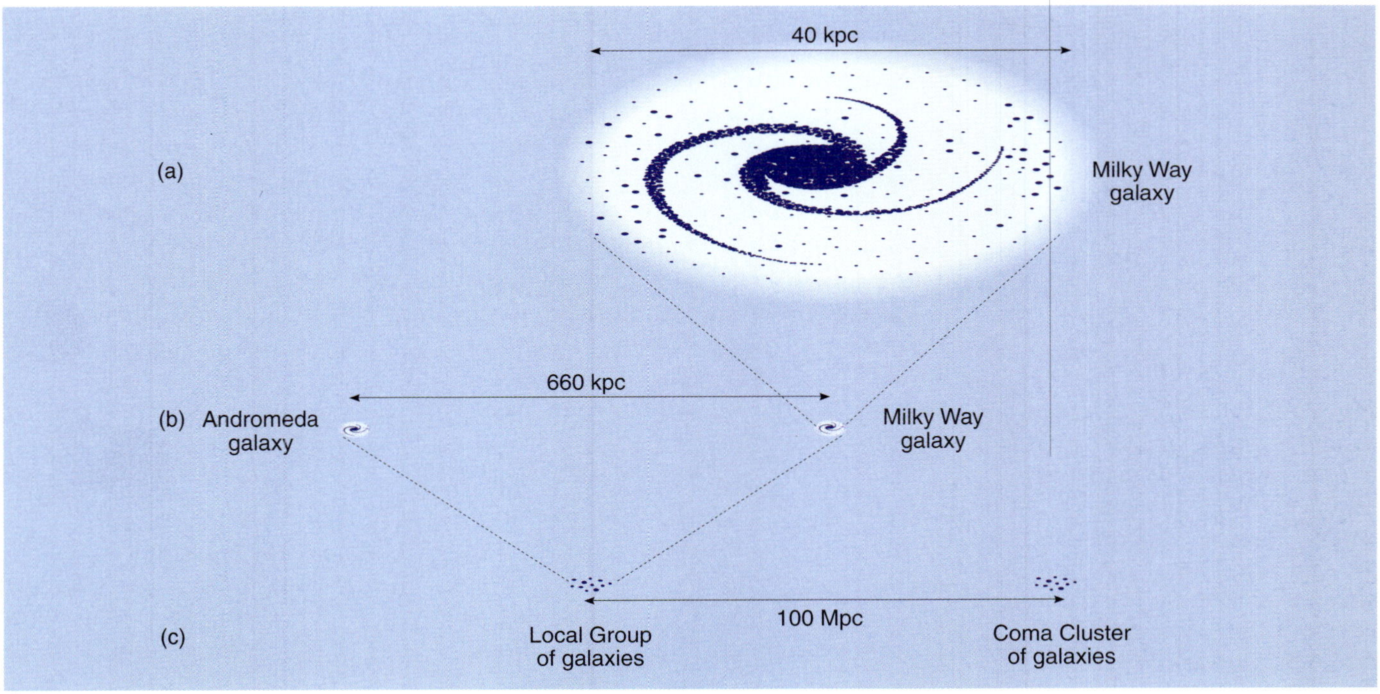

Figure 8.1 A schematic representation of distance scales in the Universe. (a) Our own galaxy, the Milky Way, is about 40 kpc in diameter. (b) The distance from our own galaxy to the Andromeda galaxy is about 660 kpc, which is around 16 times the diameter of the Milky Way. (c) Our own galaxy and the Andromeda galaxy are both part of the so-called Local Group of galaxies. The distance from the Local Group to the Coma Cluster of galaxies is about 100 Mpc, which means that the Coma Cluster is around 150 times further away from us than is the Andromeda galaxy.

Question 8.1

A cluster of galaxies is said to be at a distance of 200 Mpc from the Earth.

(a) What is this distance in kilometres?

(b) How long would it take a beam of light to travel from the cluster to the Earth?

Measuring the distance to galaxies is a difficult business though. Obviously it is not possible to simply ‘take a journey’ to the galaxy in question and measure the distance that way. As noted earlier, all that we have to go on is the light from the galaxy and, in particular, how bright the galaxy appears to be. However, in certain circumstances this turns out to be enough information to work out the distance to a galaxy. The reason is that the **brightness** of a galaxy depends on two things: the amount of light it emits (this is called the **luminosity** of the galaxy – it is a measure of the power of the galaxy in watts), and how far away it is. To appreciate this, consider the following ‘thought experiment’.

■ Imagine that you have two identical lamps – both have the same luminosity, say 100 watts, and both emit their light in all directions. You switch them on, then place one of them 100 m away from you, and the other only 20 m away. In the dark, how can you tell which of the two lamps is nearer to you?

□ The nearer lamp will appear to be brighter than the one that is more distant. So, even though the two lamps have the same luminosity, they have different brightnesses because they are at different distances away from you.

To appreciate the relationship between the luminosity and brightness of an object, look at the situation shown in Figure 8.2. This shows two galaxies labelled A and B, which have the same luminosity, but are at different distances from the Earth. As the light from each galaxy travels out into space in all directions, so the light spreads out over the surfaces of imaginary spheres, centred on the galaxy. By the time it reaches the Earth, the light from the more distant galaxy is spread out over a larger sphere than the light from the nearer galaxy. So the more distant galaxy will appear less bright than the nearby galaxy since its light is spread out over a bigger area.

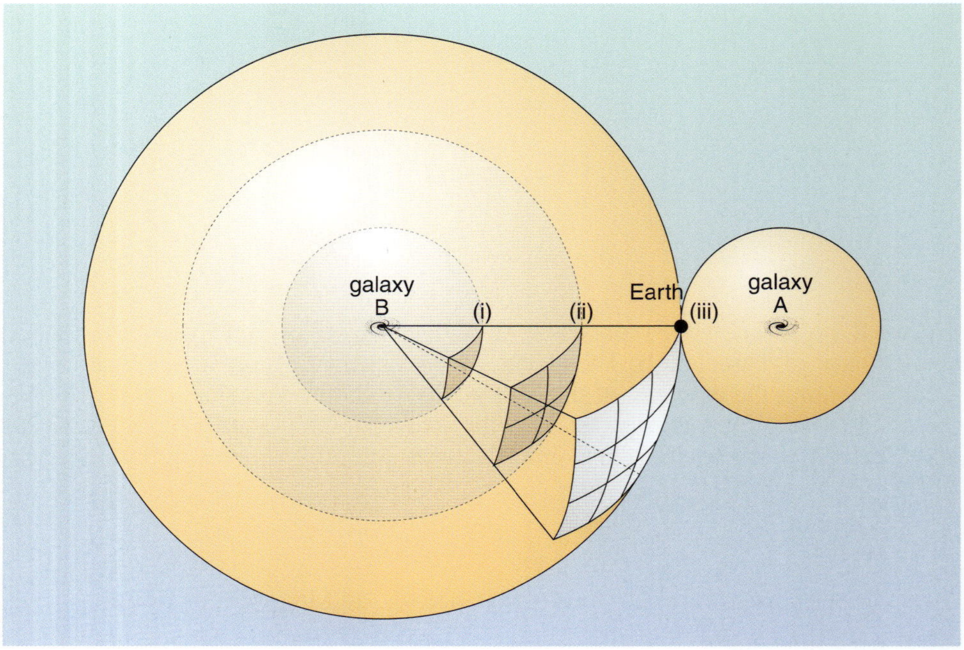

Figure 8.2 The light from distant galaxies spreads out over the surfaces of imaginary spheres centred on each galaxy. Galaxies A and B have the same luminosity but are situated at different distances from the Earth. For galaxy B, three spherical surfaces (i, ii and iii) are shown at regularly increasing distances away from the galaxy. Galaxy A is closer than galaxy B, its light is spread out over a smaller sphere by the time it reaches the Earth, and so it appears brighter.

In fact, it is convenient to *define* the brightness of a galaxy to be equal to its luminosity divided by the surface area of an imaginary sphere whose radius is equal to the distance from the Earth to the galaxy. The surface area of a sphere is given by $4\pi r^2$, where r is the radius of the sphere. So, if the radius is doubled, the surface area increases by a factor of four, and if the radius increases by a factor of three, the surface area increases by a factor of nine.

The word 'flux' is often used instead of 'brightness', so using the symbol F to represent the brightness of a galaxy (or the flux of light received from it) you can write:

$$F = \frac{L}{4\pi r^2} \tag{8.1}$$

where L is the galaxy's luminosity and r is its distance away from us. The SI unit for flux or brightness is therefore watts per square metre (W m^{-2}). This relationship between brightness and luminosity is said to be an **inverse square law**. Any relationship where one quantity *decreases* as the square of another quantity *increases* can be classified in this way. You will meet two other examples of inverse square laws later in the book.

As has been repeatedly emphasised, when observing a distant galaxy, cosmologists can measure its brightness, but if the luminosity of the galaxy is known, then the distance to it can be calculated.

■ If galaxy A in Figure 8.2 has a luminosity L of 1.0×10^{42} W, and its brightness F is measured as 9.0×10^{-10} W m^{-2}, how far away is the galaxy?

□ First, Equation 8.1 must be rearranged to make r the subject:

$$r^2 = \frac{L}{4\pi F}$$

and taking the square root of both sides:

$$r = \sqrt{\frac{L}{4\pi F}}$$

So in this case:

$$r = \sqrt{\frac{1.0 \times 10^{42}\ \text{W}}{4\pi \times 9.0 \times 10^{-10}\ \text{W m}^{-2}}} = \sqrt{8.8 \times 10^{49}\ \text{m}^2} = 9.4 \times 10^{24}\ \text{m (to two}$$

significant figures)

Since 1 pc is equal to 3.1×10^{16} m, there are 3.1×10^{16} metres per parsec (i.e. 3.1×10^{16} m pc^{-1}), so

$$r = \frac{9.4 \times 10^{24}\ \text{m}}{3.1 \times 10^{16}\ \text{m pc}^{-1}} = 3.0 \times 10^8\ \text{pc or 300 Mpc}$$

But how is the luminosity of a galaxy determined in the first place? One such method of determining the luminosity is as follows. Clusters of galaxies can contain anything from a few dozen to a few thousand individual galaxies. When cosmologists look at a cluster, such as the Coma Cluster of galaxies shown in Figure 8.3, they see that all the galaxies within it have different brightnesses. Now, on the scale of the Universe it is usually adequate to assume that all galaxies within any individual cluster are at about the same distance from us. (Remember that the distances between individual galaxies in a cluster – a few hundred kiloparsecs – are small when compared with the distance of the cluster from us – usually hundreds of megaparsecs.) So the variation in brightness of galaxies *within* a cluster must reflect an *intrinsic* variation in luminosity from one member galaxy to the next. The assumption that cosmologists make is that, wherever they find reasonably large clusters (say more than a hundred members), the *tenth brightest* galaxy in any one cluster has roughly the same luminosity as the tenth brightest galaxy in any other cluster. The tenth brightest is therefore assumed to be a typical galaxy for any cluster and is referred to as a **standard candle**. Although details vary depending on the region of the spectrum in which the luminosity is measured, a typical value for the luminosity of the tenth brightest galaxy in a cluster is around 10^{41} W. (For comparison, recall that a typical household light bulb has a luminosity of only 10^2 W or less.)

Figure 8.3 An image of the Coma Cluster of galaxies, covering an area of sky about one-quarter the size of the full Moon. This cluster of galaxies is situated at about 100 Mpc away, and individual members have a range of different brightnesses. In fact, virtually all of the objects in this image are galaxies in the cluster. An exception is the bright object at the top, centre which is a foreground star within our own galaxy (the 'spikes' seen here are an artefact caused by the camera).

The idea that the tenth brightest galaxy in any cluster of galaxies is a standard candle of constant luminosity has been checked by more direct means in nearby clusters where individual bright stars can be picked out within the component galaxies. For instance, there is a class of stars known as Cepheid variables, members of which vary in brightness in a periodic manner, getting brighter then fainter every few days. The period of variability is directly related to the average luminosity of the star: the more luminous the star, the longer the period. The period can be measured simply by monitoring the star's brightness and determining the duration of the cycle of variability, and this allows the average luminosity to be calculated. By comparing the calculated average luminosity to the observed average brightness, the distance to the Cepheid variable star can be calculated. This will therefore indicate the distance to the star's host galaxy, and so the cluster of galaxies within which it sits. These measurements indicate that the assumption of the tenth brightest galaxies all having roughly the same luminosity is valid for nearby clusters. Therefore it can be assumed that the luminosity of the tenth brightest galaxy in *any* cluster is the same, and so the distance to *any* cluster can be found by measuring the brightness of its tenth brightest member.

In practice, the procedure for determining the distance to a cluster is often based on comparing the brightnesses of a pair of galaxies in different clusters, which are assumed to have the same luminosity (both are tenth brightest), and where the distance to one of them is already known.

■ Suppose that Figure 8.2 shows the tenth brightest galaxies, labelled A and B, in a couple of clusters. Galaxy B is three times further away than galaxy A. How will their brightnesses compare?

□ Since galaxy B is three times further away than galaxy A, galaxy B will appear to be $3 \times 3 = 9$ times fainter than galaxy A.

■ If, instead, galaxy A were 49 times brighter than galaxy B, and situated 100 Mpc away, what would be the distance to the cluster in which galaxy B sits?

□ Galaxy A and galaxy B are both tenth brightest in their clusters so the assumption is made that both have the same luminosity. Since galaxy B is 49 times fainter than galaxy A, its light must be spread out over the surface of a sphere that has 49 times greater area than that for galaxy A. The surface area of a sphere is proportional to its radius squared, so to get a surface area that is 49 times larger, the radius of the sphere must be seven times greater (since $7^2 = 49$). So galaxy B must be seven times further away than galaxy A, i.e. at a distance of 700 Mpc.

Question 8.2

(a) Two galaxies, referred to as G1 and G2, have the *same* luminosity as each other, but whereas G2 is only 250 kpc away, G1 is 2250 kpc away. How do the relative brightnesses of the two galaxies compare?

(b) Two other galaxies, referred to as G3 and G4, have the same luminosity as each other but G3 is 36 times fainter than G4. If G3 is 1500 kpc away, how far away is G4?

8.2 The apparent speed of galaxies

This section considers how to measure the apparent speed with which a galaxy is moving with respect to us. Once again, this is based on measurements made on the light emitted by a galaxy, but this time what needs to be examined is the spectrum of the light (i.e. how it is distributed with wavelength) rather than the total amount of light emitted. It may not be immediately obvious what spectra have to do with speed measurements, but this will soon become apparent.

In a similar manner to the experiment you carried out in Book 3 (Activity 11.1), the way to measure the spectrum of light from a star or galaxy is to point a telescope at the object in question and allow the light to pass through a **diffraction grating**. In a telescope, the spectrum is usually recorded using an electronic imaging sensor called a CCD (charge-coupled device), which is similar to the imaging sensor in a home video camera, and the information is stored on a computer for processing.

When the light from stars or galaxies is spread out to form spectra (e.g. Figure 8.4a), the spectra are seen to contain many dark lines superimposed on the overall bright background. These are **absorption lines** and are due to the presence of particular types of atoms. As noted in Section 2.3, each type of atom absorbs light of particular wavelengths. The bright, continuous background in the spectrum from a star is produced by photons coming from deep in its atmosphere.

As these photons emerge through the star's cooler, outermost layers, photons with specific energies are absorbed by atoms. The absorption lines are therefore characteristic of the particular elements that are present in the outermost layers of the star.

As you saw in Book 3, Chapter 8, there are two ways of representing such a spectrum. First, a spectrum can be displayed as an image. Figure 8.4a shows such an image of a spectrum from a typical star. The intensity of the light at any point in the spectrum is represented by the brightness of the image at that point – darker shades indicate lower intensity. Second, the spectrum may be displayed as a graph, known as a spectral distribution, as shown in Figure 8.4b. This is the same spectrum as in Figure 8.4a, just displayed in the alternative representation.

When the spectra of distant galaxies are examined, similar absorption line spectra are seen to those of stars in our own galaxy. An example of a galaxy spectrum is shown in Figure 8.4c, where it may be compared with the spectrum of the typical star in Figure 8.4b. It is not surprising that the two spectra are rather similar, since the spectrum of the galaxy is simply the sum of the spectra of the billions of stars of which it is composed. Because most galaxies are far away, telescopes are unable to distinguish individual stars within them.

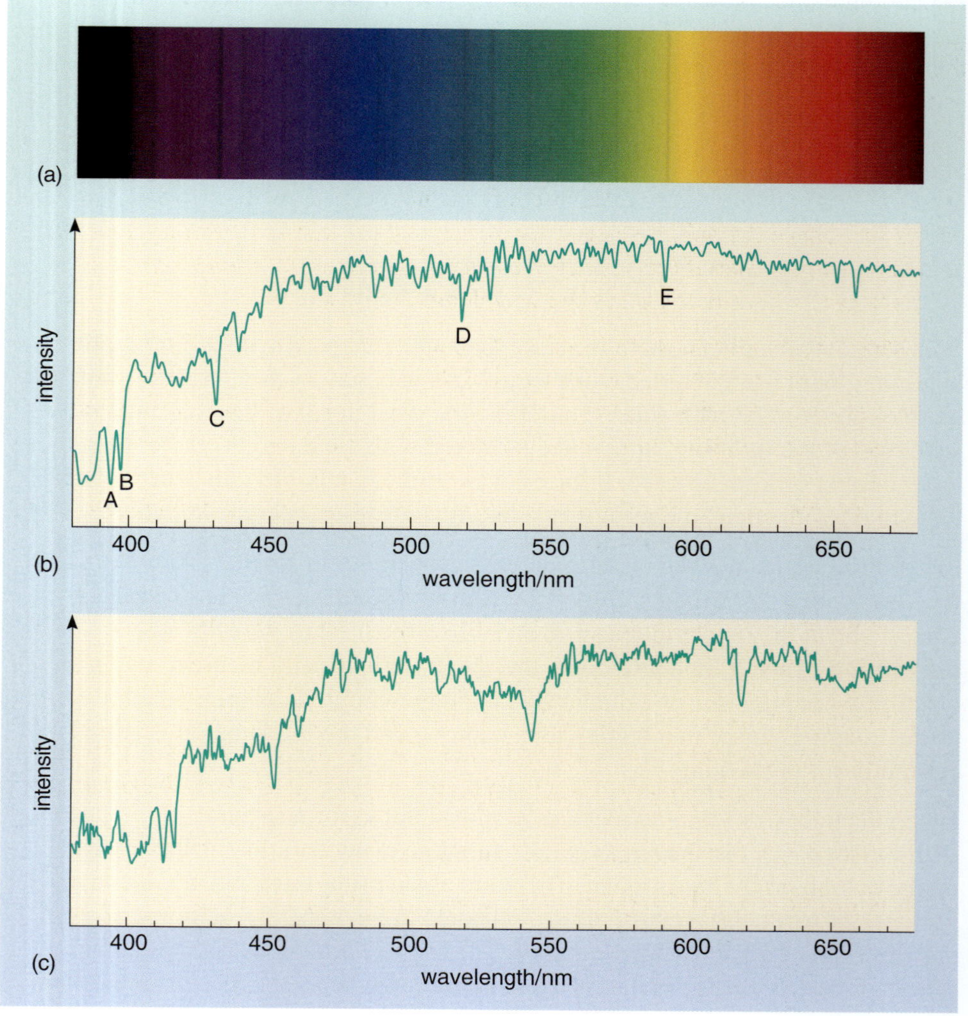

Figure 8.4 (a) and (b) show two representations of the spectrum of a star within our own galaxy. Here, violet light is displayed towards the left of the spectrum (wavelength 380 nm) and red light to the right (wavelength 680 nm). (a) An image of the diffraction pattern. (b) The spectral distribution of the same spectrum. Absorption lines are visible as relatively dark bands in (a) and as dips in the graph in (b). (c) A spectrum of a distant galaxy shown in the same representation as (b).

■ Five of the strongest absorption lines in the spectrum of the star are labelled A–E in Figure 8.4b. Can you identify this same pattern of lines in the spectrum of the galaxy (Figure 8.4c)? What do you notice about the positions of the lines in the galaxy spectrum, relative to their positions in the star spectrum?

□ The same basic *pattern* of absorption lines appears in each spectrum, but the *positions* of the lines are different. In particular, the lines in the spectrum of the galaxy (Figure 8.4c) are displaced to longer wavelengths, relative to those in the star (Figure 8.4b). For example, absorption line C appears at about 430 nm in Figure 8.4b but at about 450 nm in Figure 8.4c.

It is assumed that the absorption lines seen in the spectrum of the distant galaxy are due to the *same* elements as those seen in the spectra of the Sun and other stars within our own galaxy. The difference is that the absorption lines in the spectrum of the distant galaxy are shifted towards longer wavelengths.

Shifted wavelengths have a very natural interpretation in everyday life. The phenomenon is known as the **Doppler effect** and it is probably familiar to you in the context of sound waves, although it applies equally to any wave motion, including electromagnetic radiation such as light. The Doppler effect with sound is perhaps most noticeable when an approaching ambulance sounds its siren or as a speeding car races past. As the vehicle approaches and then recedes, apart from growing louder and then fainter, the pitch of the sound is perceived as higher when the vehicle is approaching than when it is receding. As shown in Figure 8.5, this can be understood in terms of the sound waves getting 'bunched up' in front of the vehicle as it approaches, and 'stretched out' behind the vehicle as it recedes. This happens simply because the vehicle moves between the time it emits a particular crest of the sound wave and when it emits the next crest. The bunching up of the waves in front of the vehicle causes the wavelength of the sound reaching your ears to be shorter than if the vehicle were stationary, and the stretching out of the waves behind the vehicle causes the wavelength to be longer.

A shorter wavelength of sound implies a higher pitch of the sound. So, as the vehicle approaches, you hear a higher pitch than if the vehicle were stationary. Conversely, as it recedes from you, the pitch will be lower than if the vehicle were stationary. A similar effect is observed with light.

■ The wavelength of sound waves may be appreciated by the pitch of the sound perceived by the human ear. How does the wavelength of light manifest itself to human senses?

□ The wavelength of a light wave is perceived by the human eye as the colour of the light.

A shift in the colour to longer wavelengths (towards the red end of the spectrum) is an indication of motion away from the observer, a shift in the colour to shorter wavelengths (towards the blue) is an indication of motion towards the observer.

■ Which way is the spectrum shifted in Figure 8.4c?

□ It is shifted towards longer wavelengths, i.e. towards the red.

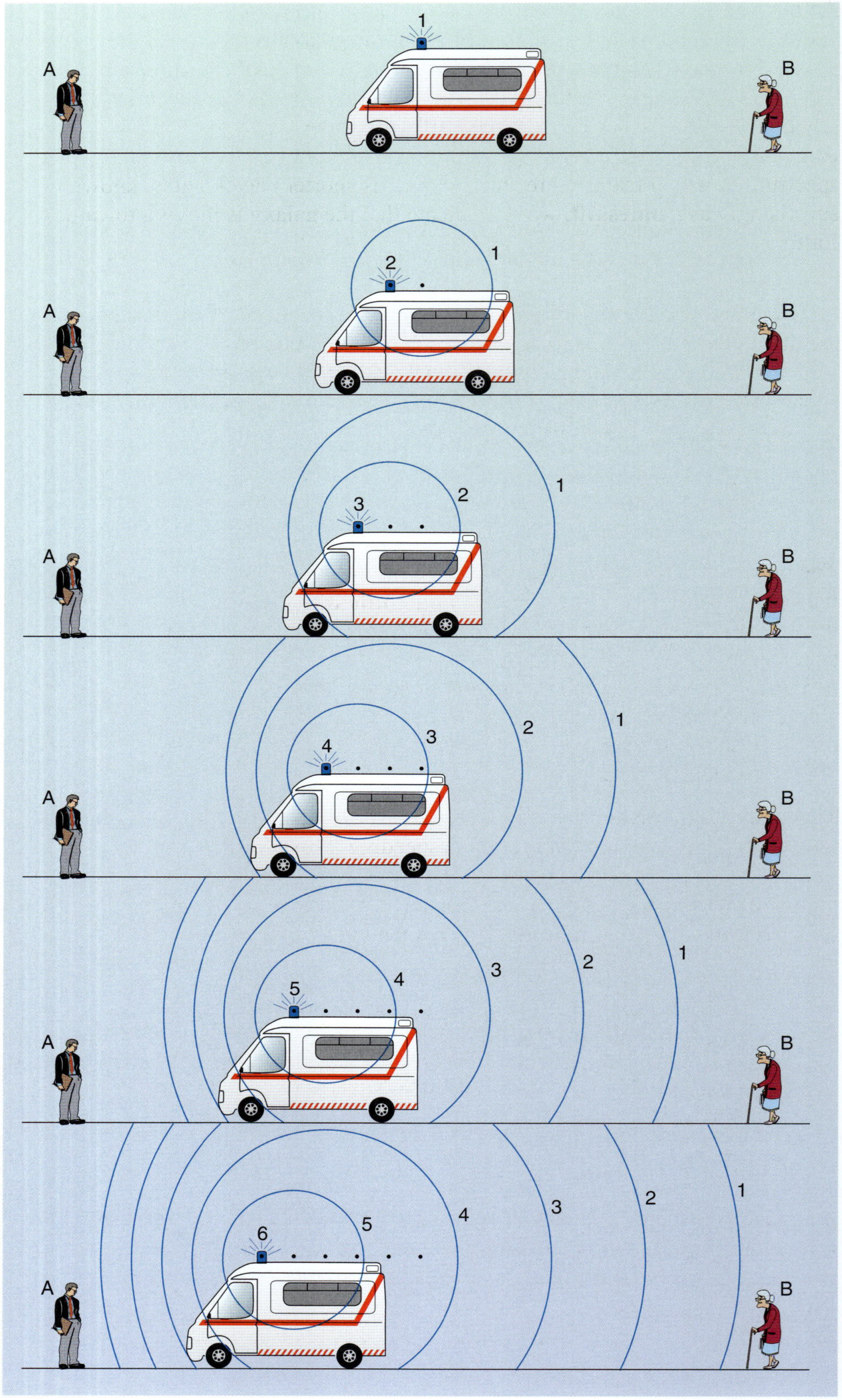

Figure 8.5 A demonstration of the Doppler effect with sound (not to scale). The ambulance sounds its siren as it moves to the left towards observer A and away from observer B on the right. Six successive time intervals are shown, running from top to bottom in the six sketches. The circles emanating from the ambulance represent successive crests of the sound wave emitted by the siren. The dots represent the centres of these successive circles. You may like to think of the circles as being similar to the ripples produced when a stone is dropped into a pond – the wave crests spread out from the centre just as shown here. The wavelength is the distance between any two successive crests at any point. By the time the second wave crest is emitted, the ambulance has caught up slightly with the first wave crest. By the time the third wave crest is emitted, the ambulance has caught up with the second wave crest, and so on. The consequence is that a person at A will perceive a sound wave with a shorter wavelength and higher pitch than that emitted by the siren when at rest, whilst a person at B will perceive a longer wavelength and lower pitch.

Astronomers say that the galaxy spectrum shown in Figure 8.5c displays a **redshift**, and an interpretation of this is that the galaxy is moving away from the Earth. Spectral lines produced by atoms in the galaxy have the same wavelength as lines produced in similar atomic processes in an Earth-based laboratory. But if the galaxy is moving away from the Earth, the wavelengths of its lines observed on Earth will be shifted towards the red. By exactly the same reasoning, a spectrum in which features are shifted towards shorter wavelengths, known not surprisingly as a **blueshift**, would indicate that the galaxy is moving towards the Earth.

The redshift, or blueshift, is defined as the *change* in wavelength divided by the *original* wavelength. Using the symbol z to represent the redshift or blueshift, λ_0 to represent the original wavelength, and $\Delta\lambda$ to represent the change in wavelength (i.e. observed wavelength minus original wavelength), this may be expressed as:

$$z = \frac{\Delta\lambda}{\lambda_0} \qquad (8.2)$$

The original wavelength is assumed to be that which would be produced by the same type of atoms in a laboratory on Earth. It is also sometimes known as the **rest wavelength** since it is the wavelength that would be observed from a stationary source. To identify which types of atoms produced the absorption lines in a galaxy spectrum, a certain amount of pattern matching is required to compare whole series of lines rather than just one or two individual lines. The same value of the redshift must apply to *all* lines in a spectrum of a certain galaxy, whatever their individual wavelengths.

The natural interpretation of observing a redshift in the spectrum of a galaxy would be that the galaxy in question is speeding away from us. However, as you will see shortly, this naive interpretation turns out to be not quite the correct picture of what is happening in the Universe. Nonetheless, it is often convenient to convert the measured redshift into an apparent speed of motion.

For the speeds that will be considered here, the apparent speed of motion of the galaxy is equal to the redshift multiplied by the speed of light.

Using the symbol v to represent the apparent speed of the galaxy, this may be expressed as

$$v = z \times c \qquad (8.3)$$

Remember that the speed of light, c, is about $3.0 \times 10^8\ \text{m s}^{-1}$ or $3.0 \times 10^5\ \text{km s}^{-1}$. If the wavelength of an absorption line in the spectrum of a galaxy is measured and compared with the wavelength of the same spectral line, as measured in a laboratory, the redshift of the galaxy can be calculated. This can then be converted into an apparent speed of recession (i.e. motion away from us).

Worked Example 8.1

As an illustration of the use of shifted spectral lines to determine apparent speeds of motion, calculate how fast the galaxy whose spectrum is shown in Figure 8.4c is apparently moving. (The spectral lines in the spectrum of the star (Figure 8.4b) occur at the *same* wavelengths as in spectra produced in the laboratory. It may therefore be assumed that the star is 'at rest' with respect to the Earth.)

The absorption line labelled C in the spectrum of the star occurs at a wavelength of 431 nm. The corresponding line in the spectrum of the galaxy (Figure 8.4c) occurs at a wavelength of 452 nm. The rest wavelength is therefore 431 nm, and the change in wavelength of this line when observed in the galaxy spectrum is (452 – 431) nm = 21 nm. Since the observed wavelength is longer than the rest wavelength, this galaxy exhibits a redshift and the galaxy is receding from us. Using Equation 8.2, the redshift is calculated as:

$$z = \frac{21\ \text{nm}}{431\ \text{nm}} = 0.049$$

Using Equation 8.3, the apparent speed that this redshift corresponds to is then simply $(0.049 \times 3.0 \times 10^5)$ km s^{-1} or 1.5×10^4 km s^{-1} (to 2 significant figures).

Question 8.3

A fast jet aircraft has a red light on its tail that emits light with a wavelength of 656 nm. The aircraft travels away from you at a speed of 0.60 km s^{-1} (about twice the speed of sound). You may assume that the speed of light is 3.0×10^5 km s^{-1}.

(a) Use Equation 8.3 to calculate the redshift of the light reaching you from the tail of the aircraft.

(b) Use Equation 8.2 to calculate the shift in wavelength of the light that you observe.

Question 8.4

The spectrum of light from a distant galaxy contains absorption lines that are identified as being due to hydrogen atoms. A particular line is observed at a wavelength of 500.7 nm, compared with the wavelength of 486.1 nm that would be produced by a source at rest in the laboratory.

(a) Is the galaxy receding from or approaching towards the Earth?

(b) What is the value of the redshift or blueshift for the galaxy?

(c) What is the apparent speed of the galaxy with respect to the Earth? (You may assume that the speed of light is 3.0×10^5 km s^{-1}.)

As noted in Section 8.1, our galaxy, the Milky Way, is one member of a small family of nearby galaxies known as the Local Group. Within the Local Group, a variety of redshifts and blueshifts are observed. This indicates that, in our local neighbourhood, the galaxies are milling around in a fairly random manner.

However, if galaxies and clusters of galaxies that are more distant than the Local Group are observed, a remarkable effect is seen: *all* the galaxies exhibit redshifts; *none* exhibit blueshifts.

■ Why is this remarkable?

□ It is remarkable because it shows that *all* clusters of galaxies in the Universe are receding from our own Local Group of galaxies!

As you will see soon (Section 8.5), this is *not* proof that we live at the centre of the Universe.

8.3 The Hubble relationship

You've seen that the apparent speeds of distant galaxies can be determined by using redshift measurements, and that their distances may be calculated by comparing the brightnesses of galaxies with their luminosities. When these results are examined for a large number of clusters of galaxies, a quite startling relationship becomes clear:

> The further away a galaxy is, the larger its apparent speed of motion away from us.

The first person to point this out was the American astronomer Edwin Hubble (Figure 8.6) in 1929. The Hubble relationship may be expressed by the simple statement:

> The apparent speed with which a galaxy is moving away from us is equal to a number, known as the **Hubble constant**, multiplied by its distance away. Using the symbol H_0 to represent the Hubble constant, and r to represent the distance to the galaxy, the Hubble relationship may be expressed as:
>
> $$v = H_0 \times r \qquad (8.4)$$

Figure 8.6 Edwin Powell Hubble (1889–1953), an American astronomer who was the first to provide definite proof that the objects now known as galaxies lie far beyond the Milky Way. He established the speed–distance relationship for distant galaxies, and the constant relating the two is known as the Hubble constant. He also produced a classification scheme for galaxies – dividing them into ellipticals, spirals and barred spirals – based upon their shape. He worked for most of his career at the Mount Wilson Observatory in California. The Hubble Space Telescope, launched by NASA in 1990, is named in his honour.

Since the apparent speed of a galaxy is typically measured in units of km s^{-1}, and the distance to a galaxy is usually measured in units of Mpc, a sensible unit for the Hubble constant can be found as follows. Rearranging Equation 8.4 gives:

$$H_0 = \frac{v}{r}$$

So the Hubble constant is equal to speed divided by distance. Hence in this case the unit is:

$$\frac{\text{km s}^{-1}}{\text{Mpc}}$$

A sensible unit for the Hubble constant is therefore km s^{-1} Mpc^{-1} (i.e. kilometres per second per megaparsec). Note that the Hubble 'constant' is in fact rather poorly named, as it is certainly not constant, but changes in value as the Universe ages. For this reason it is sometimes referred to as the *Hubble parameter*, and the subscript '0' on H_0 is a reminder that this refers to the value that is measured at the present time.

Activity 8.1 The virtual telescope

We expect this activity will take you approximately 90 minutes.

In order for you to consolidate your understanding of the measurements that lead to the Hubble relationship, you should now work through this computer-based activity and derive your own value for the Hubble constant.

In this virtual experiment, you will have control of a powerful telescope with which you can measure the redshifts and brightnesses of galaxies within several clusters. By converting the redshifts into apparent speeds and determining distances from the brightness measurements, you will be able to determine a value for the Hubble constant. As you will see later, the Hubble constant also gives a measure of the age of the Universe.

You should now work through the computer-based activity *The Virtual Telescope*. When you have completed it, look at the comments on this activity at the end of this book.

The 'virtual telescope' only allowed you to make measurements of a maximum of eight clusters of galaxies. Despite this, you should have come up with a value for the Hubble constant that is similar to the most accurate determinations made today. Figure 8.7 shows a graph of the apparent speed and distance for several hundred clusters of galaxies. As with your own measurements, a straight line has been drawn through the data points. The slope of this line tells you the *rate* at which apparent speed increases with distance. A steeper slope means that apparent speeds increase rapidly as we observe ever more distant galaxies; a shallower slope implies that apparent speeds increase only gradually for distant galaxies. The actual value of the slope of this line can be measured as the overall increase in apparent speed (3.5×10^4 km s^{-1}) divided by the overall increase in distance (500 Mpc). It gives the value of the Hubble constant as 70 km s^{-1} Mpc^{-1}. What does the value for the Hubble constant imply? Well, look at its units: kilometres per second per megaparsec. This means that for every megaparsec of distance out into the Universe, the galaxies and clusters appear to be moving about 70 km s^{-1} faster.

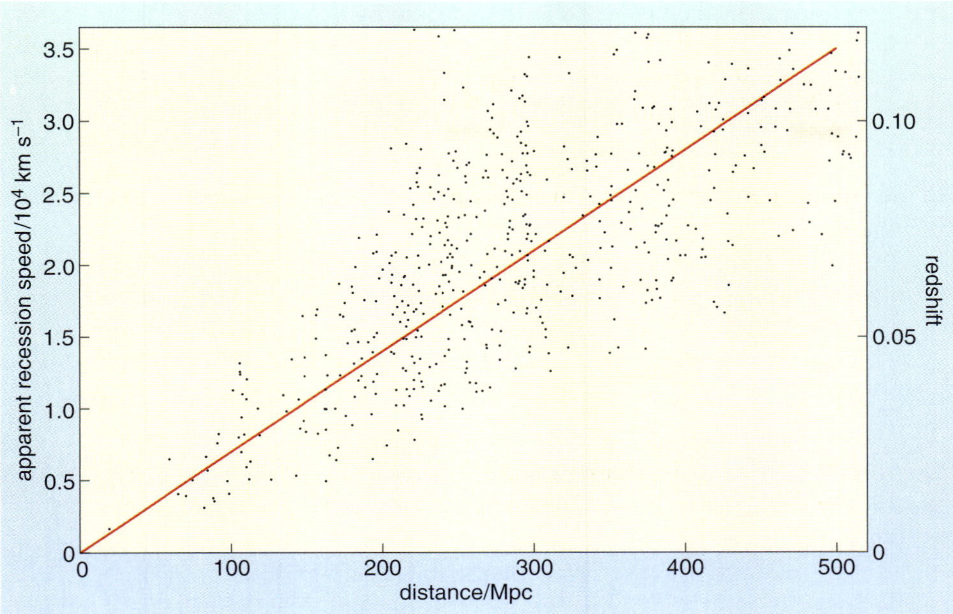

Figure 8.7 The Hubble relationship between the apparent speed of a galaxy and its distance from us, illustrated with a sample of several hundred clusters of galaxies. Note the redshift scale on the right of the graph. The large scatter that is seen in these points reflects the difficulty in determining the redshift and distance of these clusters. The redshifts (and the apparent speeds calculated from the redshifts) are typically measured to a precision of 10%, whilst the distances are only precise to about a factor of two in some cases.

Question 8.5

Assuming that the Hubble constant is 70 km s^{-1} Mpc^{-1} and the speed of light is 3.0×10^5 km s^{-1}:

(a) What would be the redshift of a galaxy at a distance of 300 Mpc from the Local Group?

(b) What would be the distance to a galaxy that has a redshift of 0.056?

(*Hint*: calculate the apparent speed of the galaxy first.)

At the time of writing (late 2007), the most precise measurement of the Hubble constant puts its value at 70.4 km s^{-1} Mpc^{-1} with an uncertainty of about ±1.5 km s^{-1} Mpc^{-1} (i.e. the true value probably lies between about 69 km s^{-1} Mpc^{-1} and 72 km s^{-1} Mpc^{-1}). By the time you read this, it is possible that an even more precise value for the Hubble constant will have been obtained using one of the new generation of powerful telescopes.

In fact, nowadays, the Hubble relationship is used to *determine* how far away certain objects are. The most distant objects that astronomers observe are quasars, which were mentioned in Chapter 1. Quasars are bright, point-like sources of light that lie at the heart of very distant galaxies. They are believed to be super-massive **black holes** (10^8 or 10^9 times the mass of the Sun) that are swallowing whole stars that stray too close to them. In the final moments before the stars are engulfed, they are ripped apart and form a swirling disk of material around the black hole. Frictional forces within this disk cause it to get immensely hot and radiate huge amounts of electromagnetic radiation. This is what makes quasars so luminous that they can be seen such a long way away. Astronomers can measure the redshift of a quasar from the spectral lines it emits, infer an apparent speed of motion away from us, and then use Equation 8.4 to determine how far away it is. Note, however, that there are a few complications when dealing with 'distances' and 'speeds' of objects that are so far away, which are addressed in the following sections.

8.4 Expanding space

You have seen how cosmologists obtain the observational data on which the Hubble relationship is based. You will now look at how it may be interpreted, and then consider the consequences of this interpretation for the properties of the Universe in the distant past.

Wherever cosmologists look, they see galaxies apparently rushing away from the Local Group and the further away the galaxies are, the faster they appear to move. The interpretation of this is that space *itself* is expanding uniformly, and the same behaviour would be observed *wherever* in the Universe you happened to be.

Here then is the reason why care has been taken up to now to refer to the 'apparent speed' of motion of galaxies away from us. Galaxies are not really rushing away from us at ever greater speeds the further away we look; it is simply that the space *in between* us and the galaxy in question is *expanding*. This expansion of space is what gives rise to the apparent increase in speed of motion with increasing distance.

Now, it is quite difficult to appreciate this for our three-dimensional Universe, so, in order to make things simpler, you will look at a one-dimensional case, represented by the strip of elastic shown in Figure 8.8. Showing that uniform expansion of a one-dimensional universe naturally gives rise to the Hubble relationship will hopefully make the idea easier to carry over to the real, three-dimensional case.

Five time steps are shown in Figure 8.8, each one a billion years after the last, with the third time step identified as 'today'. Moving from one time interval to the next, space (the strip of elastic) expands uniformly, and all the clusters of galaxies (the buttons) get further apart. Imagine that cluster A is the Local Group of galaxies that contains the Milky Way and us. The separation of clusters A and B is 32 Mpc in the first image (represented by 32 mm on Figure 8.8), 36 Mpc in the second image, 40 Mpc in the third image, and so on. (Check these distances by measuring the separations of the 'clusters of galaxies' on Figure 8.8 if you wish.) So clusters A and B are receding from each other by 4 Mpc every billion years.

Similarly, to an astronomer in cluster A, cluster C appears at distances of 40 Mpc, 45 Mpc, 50 Mpc and so on, moving down the sequence. So cluster C is getting further away by 5 Mpc every billion years. This is a faster apparent speed than that for cluster B, and is merely a consequence of it being further away. There is more space between A and C than there is between A and B, so the expansion is greater, and the apparent speed of separation is larger. To an astronomer in cluster A, cluster D appears at distances of 24 Mpc, 27 Mpc, 30 Mpc and so on, moving down the sequence. So cluster D is getting further away by 3 Mpc every billion years – that is a slower rate than both clusters B and C since it is nearer. There is less space between A and D than between either A and C or A and B, so the expansion is less, and the apparent speed of separation is smaller.

The model works well for measurements from cluster A, which you can imagine to be the Local Group of galaxies, but how do things look to an alien astronomer in cluster B? Well, to it (the alien), cluster A is clearly receding at a rate of 4 Mpc every billion years, but cluster C is receding at only 1 Mpc every billion years. Well, that is fine because cluster C is much closer than cluster A and so has a smaller apparent speed. Similarly, cluster D is receding from cluster B at 7 Mpc every billion years. And that too is just what would be expected, since cluster D is further from cluster B than either A or C.

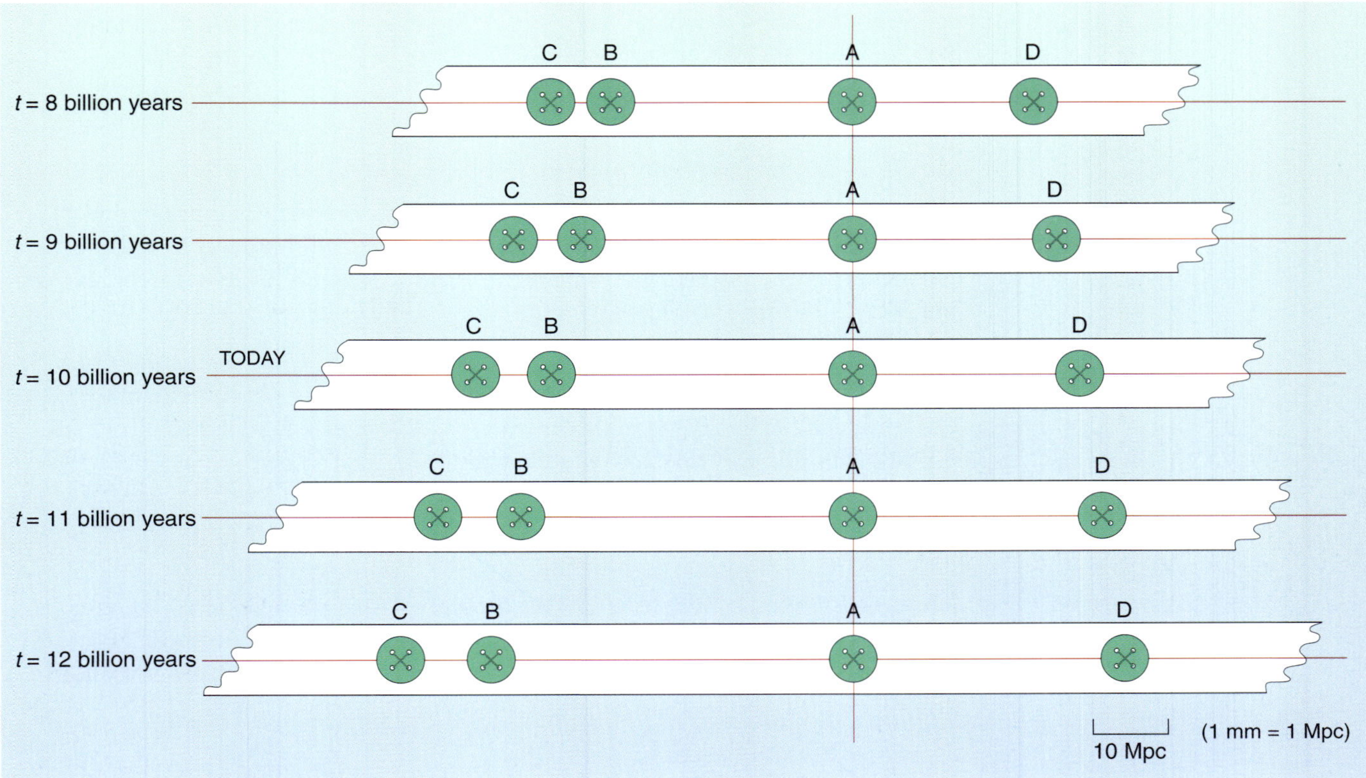

Figure 8.8 A model for the uniform expansion of part of a one-dimensional universe, shown at five time intervals, each one a billion years after the last. The strip of elastic represents space and the buttons sewn onto it, labelled A to D, represent clusters of galaxies. As space expands uniformly so the clusters of galaxies all recede from each other. As explained in the text, the apparent speed at which they recede increases with increasing separation from the point of measurement. The scale of this figure is 1 mm to 1 Mpc. In this model Universe, the current time (i.e. 'today') is identified as 10 billion years for convenience.

Wherever you happen to be in this one-dimensional universe, clusters that are further away from you recede at larger and larger apparent speeds, and this apparent speed depends on the distance, just as described by the Hubble relationship. Furthermore, in the example here, the relative apparent speed between any two clusters remains *constant* at all times. The following example will help you to appreciate the relationships between apparent speed, redshift, and the Hubble constant for the one-dimensional universe in Figure 8.8.

Worked Example 8.2

(a) Imagine that you are an alien astronomer living on a planet, orbiting a star, in a galaxy in cluster D, making observations of galaxies in clusters A, B and C in Figure 8.8. Calculate the apparent speeds of separation, convert these to km s^{-1} and then work out the redshifts of clusters A, B and C. (The speed of light is 3.0×10^5 km s^{-1}, 1 Mpc is equivalent to 3.1×10^{19} km, and 1 year is equivalent to 3.2×10^7 seconds, all to two significant figures.)

(b) Using any of the three apparent speeds you calculated in part (a), what is the value of the Hubble constant at the time identified as 'today' in this one-dimensional universe? Express your answer in units of km s^{-1} Mpc^{-1}.

(c) By carefully considering Figure 8.8, how does the value of the Hubble constant differ in the past and in the future of this one-dimensional universe from the value calculated in part (b)?

(a) The apparent speeds of recession of clusters A, B and C as measured from cluster D are 3.0 Mpc per billion years, 7.0 Mpc per billion years and 8.0 Mpc per billion years, respectively. The next step is to convert these apparent speeds into km s^{-1} because you need to compare them with the speed of light which you have in km s^{-1}.

1 Mpc is equivalent to 3.1×10^{19} km and one billion years is equivalent to (10^9 years) × (3.2×10^7 seconds per year) = 3.2×10^{16} s. So a speed of 1.0 Mpc per billion years is roughly equal to:

$$\frac{3.1 \times 10^{19}\ \text{km}}{3.2 \times 10^{16}\ \text{s}}$$

which is 1000 km s^{-1} (to 1 significant figure). This is a convenient conversion. The apparent speeds of the three clusters with respect to cluster D are therefore 3000 km s^{-1}, 7000 km s^{-1} and 8000 km s^{-1} respectively (to 1 significant figure).

Using a rearrangement of Equation 8.3, $z = \frac{v}{c}$, the redshifts of the three clusters as measured from cluster D, are:

$$\text{redshift A} = \frac{3000\ \text{km s}^{-1}}{3.0 \times 10^5\ \text{km s}^{-1}} = 0.01$$

$$\text{redshift B} = \frac{7000\ \text{km s}^{-1}}{3.0 \times 10^5\ \text{km s}^{-1}} = 0.02$$

$$\text{redshift C} = \frac{8000\ \text{km s}^{-1}}{3.0 \times 10^5\ \text{km s}^{-1}} = 0.03$$

(b) Using any of the three apparent speeds from part (a), the Hubble constant at the time identified as 'today' is found from rearranging Equation 8.4 as $H_0 = \frac{v}{r}$. So for the measurement of cluster C from cluster D:

$$H_0 = \frac{8000\ \text{km s}^{-1}}{80\ \text{Mpc}} = 100\ \text{km s}^{-1}\ \text{Mpc}^{-1}$$

The Hubble constant is measured with the same value (at the same time) *wherever* you happen to be in this one-dimensional universe, and this is true of the Universe that we find ourselves in too. This value of the Hubble constant is, however, somewhat higher than the current value in our Universe.

(c) At times in the past of this one-dimensional universe, clusters of galaxies were closer together, but moving apart at the same constant apparent speed. So the Hubble constant at earlier times was *larger* than it is at the time identified as 'today'. Conversely, at times in the future of this one-dimensional universe, clusters of galaxies will be further apart, but still moving at the same constant apparent speed. So the Hubble constant at later times will be *smaller* than it is at the time identified as 'today'.

In order to appreciate this more quantitatively, consider the value of the Hubble constant at the time indicated as 9 billion years. The apparent speed of, say, cluster C with respect to cluster D is still 8000 km s^{-1}, as calculated in part (a). This value is the same at *whatever time* you choose to measure it. The two clusters are 72 Mpc apart at this time. So the Hubble constant at this time is:

$$H_0 = \frac{8000\ \text{km s}^{-1}}{72\ \text{Mpc}} = 110\ \text{km s}^{-1}\ \text{Mpc}^{-1}$$

which is larger than it is at the time identified as 'today'.

Conversely, at the time indicated as 11 billion years, clusters C and D are 88 Mpc apart, so the Hubble constant at this time is:

$$H_0 = \frac{8000\ \text{km s}^{-1}}{88\ \text{Mpc}} = 90\ \text{km s}^{-1}\ \text{Mpc}^{-1}$$

which is smaller than it is at the time identified as 'today'.

Notice that in this universe, even though any particular two clusters of galaxies continue travelling apart with the same apparent speed at all times, the rate of expansion of the intervening space becomes progressively smaller. The Hubble constant quantifies the expansion rate of the universe, and clearly the expansion rate of this one-dimensional universe is slowing down. In our Universe too, the Hubble constant varies with time.

What was demonstrated in one-dimension with Figure 8.8 is also true in the real three-dimensional Universe in which we live; it is just a little harder to visualise, and no attempt will be made to do so here.

Having interpreted the Hubble relationship to mean that space is expanding, you can now look at the consequences of this phenomenon. It is believed that the amount of matter in the Universe is constant. So, since the separation between distant galaxies is continually increasing, this implies that the mean density of the Universe – the mass per unit volume – is continually falling.

In other words, in the distant past the Universe was very dense, whereas now, the mean density is rather low. This is the first important piece of evidence about the conditions that prevailed in the early Universe.

8.5 Strange ideas

When talking about the overall structure of the Universe, some rather awkward questions often arise. In the discussion that follows, these will be addressed in the hope that they will answer any questions you have about the behaviour of our expanding Universe. Before starting though, you should be warned that you will be required to put aside some apparently rational notions of reality and accept a few ideas that may at first seem rather strange. As you will see, it is not only artists and poets who need fertile and wide-ranging imaginations – such characteristics are equally useful for cosmologists. Although some of these ideas require you to put aside common sense, be assured that the mathematical models describing the Universe are based firmly on Einstein's theory of general relativity (Chapter 13) and are unambiguous.

8.5.1 Limitations

The first complication is that the Hubble 'constant' was larger in the distant past than it is today and, as noted in Section 8.3, for this reason it is often referred to as the Hubble parameter. There are two reasons for this. First, as you saw in the one-dimensional universe analogy (Figure 8.8), even if galaxies keep moving apart at a constant rate, the Hubble constant decreases as time progresses, simply because it is equal to apparent speed divided by separation and the separations keep getting larger. Second, the expansion rate of the Universe may have slowed down or speeded up at different times in its history and this will further change the value of the Hubble constant as time progresses. As a consequence, the distance scale of the Universe in the past is also uncertain.

Also, a word of caution is called for concerning the relationships between redshift and apparent speed that were developed earlier. Although Equations 8.3 and 8.4 are adequate for the present discussion, it would be wrong to assume that they apply in every situation. In fact, they are only true when the apparent speeds and distances involved are less than about 10% of the speed of light and 500 Mpc, respectively. At high apparent speeds, the physical meaning of 'distance' in the Universe needs to be considered carefully. The light that we now see from rapidly receding galaxies was emitted by them when the Universe was much younger than it is now. In the time it has taken that light to reach us, the Universe has expanded and so distances between galaxies have changed. Therefore, care has to be taken in interpreting the distances and apparent speeds that are measured in an expanding universe when the redshift is greater than about 0.1.

For these reasons, cosmologists do not usually refer to the apparent speed and distance of rapidly receding, very distant galaxies or quasars, but to their redshift and **look-back time**. This latter quantity is the time taken for the light emitted by a galaxy to reach us, and indicates how far back in time we are seeing. Looking at objects far away means that we are actually looking back in time, because the light that we see was emitted by the object in the distant past, and has taken a substantial amount of time to reach us.

8.5.2 Concepts

Aside from these somewhat technical issues, the idea of an expanding Universe also throws up some rather strange concepts. In response to the claim that space is expanding uniformly, many people ask (not unreasonably): 'What is the Universe expanding into?' In fact, the expansion of the Universe is interpreted very differently from an expansion of matter *into* space; rather it is interpreted as an expansion *of space itself*. Space is a property of the Universe, and matter is (more or less) fixed in a space that expands. This was illustrated by the one-dimensional universe analogy of Figure 8.8: the elastic (space) expands uniformly, but the buttons (clusters of galaxies) remain the same size and shape; they are merely carried along by the universal expansion. Similarly, the Earth is not expanding, and nor is the Solar System, the Milky Way, or even our Local Group of galaxies. These objects are all bound together by electric and gravitational forces of attraction between the atoms and molecules of which they are composed. Only beyond the scale of clusters of galaxies does the expansion win.

Although the redshift of distant galaxies has been described as being comparable to a Doppler shift, it is important to realise that there is one vital difference between a 'standard' Doppler redshift (such as that caused by speeding ambulances or the random motion of galaxies in the Local Group) and what may be called a 'cosmological' redshift . The Doppler effect is the result of the motion of an object *through* space at a certain speed; whereas cosmological redshifts are caused by the expansion *of* space itself. So, even though a distant cluster of galaxies may have an apparent speed that is, say, 88% of the speed of light, that cluster is not moving rapidly with respect to its local surroundings. In terms of the analogy in Figure 8.8, the buttons on the strip of elastic are not moving with respect to the local patch of elastic. It is the expansion of space itself that 'stretches out' the wavelength of the emitted light as it travels through space. The more space there is between the object emitting the light and the point of observation, the bigger the 'stretch', and so the larger the redshift.

Another question that many people ask is: 'Where in the Universe is the centre of this expansion?' Well, there is no centre of expansion – all space is expanding at the same rate in all directions, and the same expansion would be measured wherever you happened to be. Perhaps another analogy will help here. Figure 8.9 shows a two-dimensional universe – one step up in complexity from that shown in Figure 8.8, but still one-dimension short of the real thing. Here the Universe is represented as the two-dimensional surface of a balloon, with the buttons stuck on the surface representing the clusters of galaxies as before. It is only the *surface* of this balloon that represents space – everything inside or outside of the balloon is not part of this universe. As space expands (i.e. as the balloon is inflated), the clusters of galaxies move further apart with their apparent speeds away from each other increasing with distance just as in Figure 8.8. But the centre of expansion (the centre of the balloon) does not lie anywhere within the universe. Our own three-dimensional Universe also has no centre of expansion. The problem is that none of us can think in enough dimensions to visualise it properly. This is an important point: there is little point attempting to visualise the corresponding situation for our own three-dimensional Universe. It is almost certainly not possible to do so!

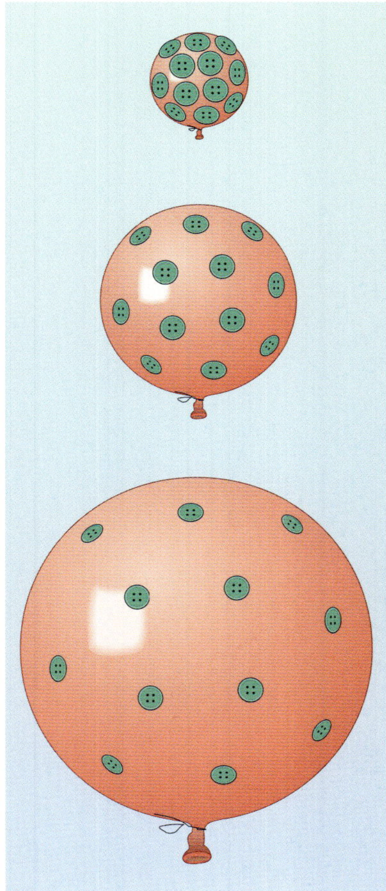

Figure 8.9 The expansion of a (finite) two-dimensional universe visualised as the surface of an inflating balloon. Notice that the centre of expansion does not lie anywhere in the two-dimensional space of this universe (the surface of the balloon). As the balloon inflates, so all the 'galaxies' (represented here by buttons stuck onto the balloon) recede from each other.

Perhaps there is a simple answer to the question: 'How big is the Universe?' Well, there is a simple answer, but it is not easy to comprehend. The Universe may be infinite in extent – *and may always have been so*. (The balloon analogy is therefore misleading in this respect, since that describes a Universe with finite size.) When it is said that space is expanding, you should *not* interpret this to mean that the overall *size* of the Universe is increasing (if it is infinite, it cannot get any bigger since infinity is the biggest possible!). Rather, you should interpret it to mean that the *separation* of large structures within the Universe is increasing; in other words, galaxies are getting further apart.

A popular misconception is to think of the Universe as originating at a 'point in space' and expanding from there. This is quite the wrong visual image, and you should try not to think in these terms. Remember, space is a property of the Universe, not something within which the Universe sits. Furthermore, the current theory for the origin of the Universe implies that the entire infinite space of the Universe, and the raw materials from which the galaxies were built, were all created at the same instant. The separations between objects increase with time, as they are carried along by the expansion of the space that was created at the instant the Universe began. Again, there is little point attempting to visualise an infinite, expanding three-dimensional Universe – it is almost certainly impossible for anyone to do so!

A final point is that there is no edge to the Universe either. Since the Universe may be infinite then, by definition, it goes on for ever and travelling in a straight line you would never reach an edge. Even if the Universe were finite in size though, you would never reach an edge. Travelling in a straight line in a finite Universe, you would eventually end up back where you started, just as an ant would crawling over the surface of the two-dimensional universe model in Figure 8.9.

The preceding few paragraphs provide a rather mind-bending excursion for most people! The problem is that we are only used to comprehending things on a much smaller scale of time and space than is necessary to grasp properly the immensity of the Universe. The ideas can be expressed mathematically, but would be an unnecessary and lengthy detour from the main story. Nevertheless, the basic ideas are not so difficult if you are prepared to discard some ideas that are 'common sense' in our everyday experience. To summarise:

> The Universe is probably infinite, with no centre and no edge, and it was probably always infinite, as far back in time as theories take us.
>
> It makes no sense to ask what is 'outside' the Universe, because space is a property of the Universe itself, and does not exist elsewhere.
>
> Space itself is expanding uniformly such that the separation between distant galaxies increases with time, and the overall density of the Universe decreases.
>
> A consequence of this uniform expansion is that the apparent speed of motion of distant galaxies increases with increasing distance from the place of measurement.

8.6 The age of the Universe

For the time being, you are going to follow this first big clue of cosmology: space is expanding. You can imagine 'running the film backwards in time' and conclude that, in the past, all the galaxies were closer together than they are now. If the assumption is made that all the galaxies we can see have been moving at their present apparent speeds since the Universe began, then the Hubble constant can be used to calculate a rough age for the Universe.

If the Hubble constant is 70 km s^{-1} Mpc^{-1}, then, from Equation 8.4, two galaxies that are currently 500 Mpc apart have an apparent speed between them of 70 km s^{-1} Mpc^{-1} × 500 Mpc = 35 000 km s^{-1}. (Remember, that is 70 km s^{-1} faster for every Mpc further apart.) So, if you imagine 'running the film backwards', these two galaxies would have been 'zero' distance apart at a certain time in the past. That time is given by:

$$t = \frac{500\ \text{Mpc}}{35\ 000\ \text{km s}^{-1}}$$

since the time for a 'journey' is simply given by the distance travelled divided by the speed. But distance divided by speed is just $\frac{1}{\text{Hubble constant}}$. So the quantity $\frac{1}{\text{Hubble constant}}$ provides a rough value for the age of the Universe, assuming that the expansion rate has been constant since time began.

So what is the age of the Universe? The value of the Hubble constant currently accepted by most cosmologists is about 70 km s^{-1} Mpc^{-1}. To work out an age for the Universe, you first need to rationalise the units here. At the moment, the Hubble constant has a unit that includes two different measures of distance: megaparsecs and kilometres. As noted earlier, 1 megaparsec is equal to about 3.1×10^{19} kilometres, so the value of the Hubble constant can be written as:

$$\text{Hubble constant} = \frac{70\ \text{km s}^{-1}\ \text{Mpc}^{-1}}{3.1 \times 10^{19}\ \text{km Mpc}^{-1}}$$
$$= 2.26 \times 10^{-18}\ \text{s}^{-1}$$

An estimate for the age of the Universe is then $\frac{1}{\text{Hubble constant}}$, so:

$$\text{estimated age} = \frac{1}{2.26 \times 10^{-18}\ \text{s}^{-1}} = 4.43 \times 10^{17}\ \text{s}$$

Since one year is equivalent to about 3.2×10^7 seconds, expressing this time as an equivalent number of years results in a value of:

$$\text{estimated age} = \frac{4.43 \times 10^{17}\ \text{s}}{3.2 \times 10^{7}\ \text{s y}^{-1}} = 1.4 \times 10^{10}\ \text{y}\ \ \text{(to two significant figures)}$$

which is 14 billion years. In practice, the true age of the Universe will be slightly different from the value given by $\frac{1}{\text{Hubble constant}}$, because the expansion of the Universe has *not* proceeded at a constant rate since time began.

8.7 Summary of Chapter 8

The intrinsic luminosity, L, and observed brightness (flux), F, of a galaxy are related by:

$$F = \frac{L}{4\pi r^2}$$

where r is the distance to the galaxy in question. The distance from the Local Group to a cluster of galaxies may be determined using the relationship between the brightness of its tenth brightest member and the standard luminosity of that galaxy. The procedure is based on the assumption that the tenth brightest galaxies in all clusters have the *same* luminosity. Distances to galaxies and clusters of galaxies are measured in units of kpc, Mpc or Gpc.

The redshift of a galaxy, z, is defined as the shift in the wavelength of a spectral line seen in its spectrum, divided by the rest wavelength of that line:

$$z = \frac{\Delta\lambda}{\lambda_0}$$

The apparent speed of motion of a distant galaxy is equal to its redshift multiplied by the speed of light: $\upsilon = z \times c$. All galaxies beyond the Local Group exhibit redshifts, so all distant galaxies are receding from us. Moreover, the apparent speed is greater for galaxies that are further away. The Hubble relationship states that the apparent speed is equal to the Hubble constant multiplied by the distance away: $\upsilon = H_0 \times r$, where the Hubble constant is measured to be around 70 km s^{-1} Mpc^{-1}.

The Hubble relationship is a consequence of the fact that space itself is expanding uniformly. Since the separation of distant objects is increasing with time, the mean density of the Universe is continually falling. Extrapolating the observed expansion back in time leads us to believe that the Universe originated about 14 billion years ago. Quasars are the most distant objects observed in the Universe. They are so far away, that we are seeing them as they were soon after the Universe formed. Because they are so distant and apparently moving so fast, normal ideas about ‘distance’ and ‘speed’ become meaningless. So the ‘apparent speed’ of a quasar is characterised by its redshift and the ‘distance away’ is characterised by its look-back time. The Universe is probably infinite in size, and always has been; it has no centre, since the same expansion would be measured from any location within it; it has no edge because space is a property of the Universe itself and does not exist elsewhere.

Chapter 9
The cooling Universe

> Time flies like an arrow. Fruit flies like a banana.
>
> Groucho Marx, American comedian (1890–1977)

Having looked closely at the expansion of the Universe, you will now examine the second major piece of evidence for an evolving Universe, namely the observation that the Universe is gradually cooling. If the Universe is cooling, then it must have a temperature, and that may seem a rather strange concept. What is the temperature of the Universe? After all, nowadays the Universe largely consists of almost empty space between the galaxies. However, space is not as empty as you might suppose. Even away from stars and galaxies, space still contains electromagnetic radiation. On average, every cubic metre of space contains about 400 million photons! These photons constitute the 'heat radiation' of the Universe and their spectrum corresponds to a particular temperature.

9.1 Black-body radiation

As you were reminded in Section 3.7, when a hot-plate on an electric cooker is heated up, it glows red-hot and radiates energy in the form of photons. These photons have a range of energies but the precise distribution of photons – the relative numbers that are emitted with any particular energy – depends on the temperature of the hotplate. As you know, the distribution of photons plotted against photon energy is simply the spectrum of the radiation. It turns out, quite reasonably, that as the temperature of an object increases, so it emits photons of progressively higher energies. There will still be a distribution of photons with different energies (i.e. a spectrum), but the *mean* photon energy will shift to higher values. An object whose emission has a mean photon energy in the blue part of the spectrum (about 3 eV) will be hotter than one whose emission has a mean photon energy in the red part of the spectrum (about 2 eV), for instance. But there is no need to restrict this relationship between photon energy and temperature to merely the visible part of the electromagnetic spectrum. At higher energies (shorter wavelengths) than blue light there are the ultraviolet and X-ray regions. Objects whose emissions have a mean photon energy ($E_{\text{ph, mean}}$) in these parts of the spectrum must be extremely hot. Conversely, at lower energies (longer wavelengths) than red light are the infrared and microwave regions. Emission that has a mean photon energy in these ranges would indicate much cooler temperatures.

In fact, this relationship may be quantified roughly as follows:

$$T \approx (3000\ \text{K eV}^{-1}) \times E_{\text{ph, mean}} \tag{9.1}$$

You met the SI unit of temperature, kelvin, symbol K, in Book 3, Section 6.4.

> As a rough rule of thumb, photons with a mean energy of about 1 eV will be produced by a body at a temperature of around 3000 K, and the mean energy of the photons emitted by other bodies is proportional to the temperature.

■ What temperature is required to produce a spectrum whose mean photon energy is 100 keV (i.e. X-rays, see Figure 2.6)?

□ 100 keV = 100 000 eV or 10^5 eV. Since photons with a mean energy of 1 eV are produced by a body at 3000 K, a temperature of around ($10^5 \times 3000$ K) = 3×10^8 K or 300 million kelvin is required in this case.

■ What is the mean energy of photons radiated by objects at a typical room temperature?

□ A typical room temperature is about 300 K. This is ten times less than the 3000 K that gives a mean photon energy of 1 eV. So the mean photon energy must be about 0.1 eV. This corresponds to the infrared part of the electromagnetic spectrum (see Figure 2.6).

The continuous spectral distributions of many objects have precisely the same shape, they are merely shifted to different energies or wavelengths. Some examples are shown in Figure 9.1 and each of these is what is known as a **black-body spectrum** or **thermal spectrum**.

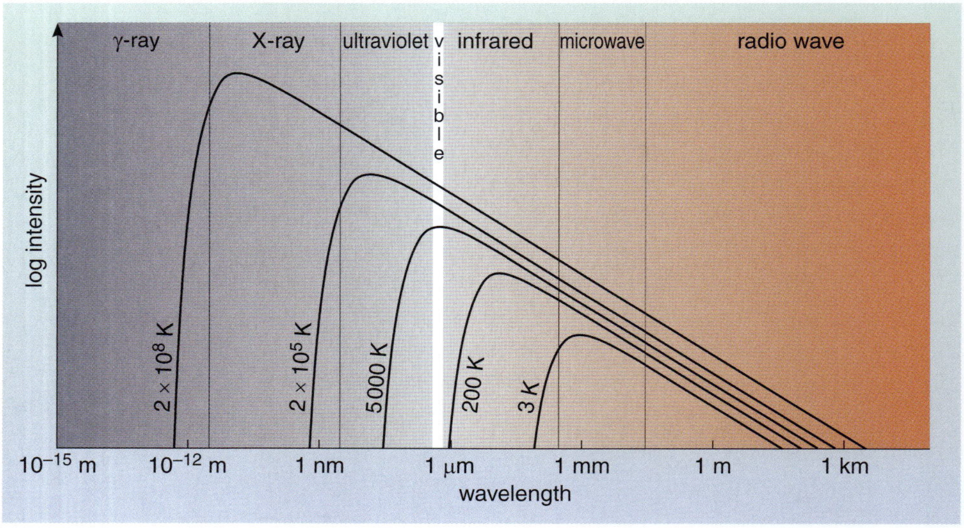

Figure 9.1 The black-body spectra emitted by objects at temperatures of 2×10^8 K, 2×10^5 K, 5000 K, 200 K and 3 K. The scale along the bottom shows the wavelength of the radiation, and you can see that hotter objects emit more of their radiation at shorter wavelengths. Notice that the wavelengths and intensities are plotted on logarithmic scales to allow a vast range to be included in one diagram.

From everyday experience, you may be aware that a black surface absorbs more radiation than a silver surface. In fact, a perfectly black surface will absorb all the radiation that falls upon it and, in a steady-state (or equilibrium) situation where it maintains a constant temperature, it will also emit all this radiation back again. (This assumes that the surface is in a vacuum and cannot exchange energy with its surroundings in any other way.) The phrase 'black body' is therefore used as a shorthand to describe any object that behaves as a perfect absorber and emitter of radiation. The crucial features of black-body spectra are that they all have the same continuous shape, they contain no emission or absorption lines, and the mean photon energy (or corresponding wavelength) depends *only* on the temperature of the object according to Equation 9.1.

The key to understanding how black-body spectra are produced is that the object and the radiation are in **thermal equilibrium**. As much radiation is being absorbed as is being emitted at every instant, and the object therefore remains at a constant temperature. It can therefore be said that the radiation also possesses this same temperature. The conditions necessary to create such a situation are generally those of high temperatures and large amounts of energy. Under such conditions, photons are rapidly absorbed and re-emitted by matter.

■ In order for a photon to be absorbed by an atom, what condition must be met by the energy levels of the atom?

□ There must be a pair of energy levels whose separation is equal to the energy of the photon concerned.

■ How could you guarantee that a photon of virtually *any* energy could be absorbed?

□ There must be a great many energy levels, very close together, and extending over a large range of energy.

It turns out that this situation will exist when atoms are ionised (i.e. when one or more electrons are removed from an atom), and also when atoms are arranged in a metal. So, in both of these cases, photons of *all* energies may be absorbed and emitted, and a continuous spectrum is produced. The continuous spectrum of the Sun and other stars (ignoring the absorption lines superimposed on top – Figure 8.4) may be approximated by black-body spectra.

Question 9.1

Use Figure 9.1 to help you answer the following questions.

(a) A domestic oven can reach a temperature of around 230 °C. What is this equivalent temperature in kelvin and in which region of the electromagnetic spectrum would the peak intensity of its black-body spectrum occur?

(b) The black-body spectrum emitted by an object has a peak intensity in the ultraviolet part of the electromagnetic spectrum. Roughly what is the temperature of the object?

9.2 The cosmic microwave background

Objects in the Universe emit electromagnetic radiation across the whole spectrum from radio waves through infrared, visible radiation, ultraviolet, and X-rays to gamma rays. The spectra from individual objects are, in some cases, characteristic of thermal processes, and so have the continuous black-body shape. If astronomers observe the Universe in the microwave part of the spectrum – that is, at wavelengths of a few millimetres – a remarkable phenomenon is observed. What they detect is a background microwave 'glow' coming from the whole Universe, and wherever they look (away from individual stars or galaxies) it has virtually the same spectral distribution and intensity. Moreover, the shape of the spectrum is that of a thermal or black-body source. This **cosmic microwave background**, or CMB, was discovered in 1964 by Arno Penzias and Robert Wilson (Figure 9.2).

Figure 9.2 Robert Wilson and Arno Penzias standing by the antenna with which they discovered the cosmic microwave background at Bell Telephone Laboratories in New Jersey, USA. They were tracking a satellite and found that their antenna picked up not only the signal from the satellite, but also some background 'noise'. What they discovered was that this noise came from all over the sky. After ruling out interference from pigeon droppings inside the horn of their detector, and finally consulting with physicists from Princeton University, they realised that they had inadvertently discovered the 'heat radiation' of the Universe. Their work played a major part in establishing the standard model of the origin and early evolution of the Universe. Penzias and Wilson were awarded the Nobel Prize for Physics in 1978 for their discovery.

Over the years since Penzias and Wilson first discovered the CMB, its spectrum and variation across the sky have been investigated and mapped with increasing accuracy. Some of the most recent, and accurate, measurements are those made using the Cosmic Background Explorer (COBE) satellite, which was launched in 1989. The average spectrum of the radiation, over the whole sky, measured by COBE is shown in Figure 9.3. The observations are an excellent fit with the theoretical curve that would be expected from a black-body source at a temperature of 2.73 K. Sometimes, this value is quoted approximately as simply 3 K (i.e. –270 °C).

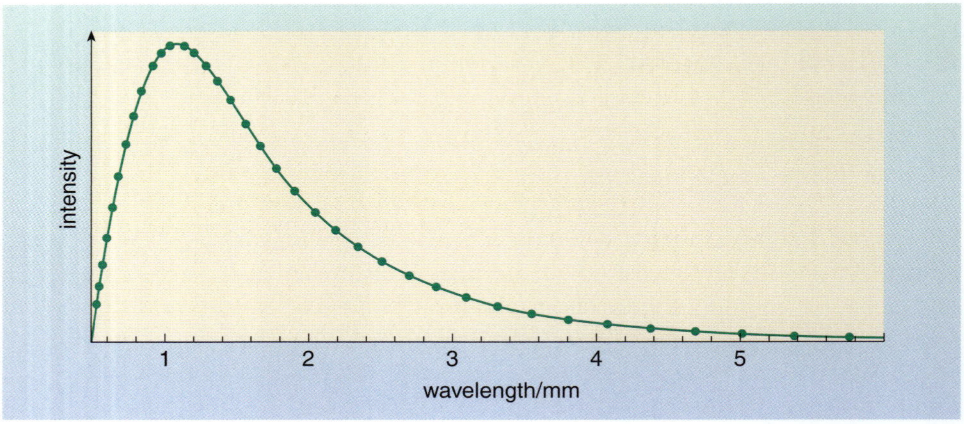

Figure 9.3 The spectrum of the cosmic microwave background radiation as measured by COBE. The points are the measured data, and the line drawn through them is a black-body spectrum corresponding to a temperature of 2.73 K. The wavelength corresponding to the mean photon energy of the spectrum is about 2 mm. The lead scientists of the COBE mission, John Mather and George Smoot, shared the 2006 Nobel Prize for Physics in recognition of their work. (Notice that whereas Figure 9.1 was drawn on a logarithmic scale in order to display the vast range of wavelengths across the electromagnetic spectrum, this graph is plotted on a simple linear scale. For this reason, the shape of the spectrum here appears different from that of the black-body spectra in Figure 9.1.)

In precise physical terms, the temperature of the CMB radiation is now (coincidentally) one hundred times colder than the normal melting temperature of ice (273 K). Yet, the spectrum has exactly the same shape as is observed in, say, a furnace whose walls are a thousand times hotter at a temperature of 3000 K, where interactions between radiation and matter rapidly create the stable distribution of photon energies necessary to produce a black-body spectrum. 3000 K can be thought of as roughly the *minimum* temperature at which atoms and radiation can interact significantly (corresponding to a mean photon energy of about 1 eV). Below this temperature, the energy levels in hydrogen atoms are simply too far apart to absorb many of the photons corresponding to a black-body spectrum.

At a temperature of only about 3 K, a steady state, or thermal equilibrium between matter and radiation in the Universe is virtually impossible to establish, since the energy of most of the photons is so very small when compared with the separation of energy levels in hydrogen atoms. So how can radiation that is now far too cold to interact with matter, to any significant extent, have acquired a thermal spectrum, when thermal spectra are generally characteristic of processes at least a thousand times hotter? Were it not for the previous observation of an expanding Universe, there would be a real puzzle here.

The solution is that the CMB radiation now seen was emitted by matter at a time in the distant past when the Universe was much hotter than it is today. As the CMB radiation travels towards us from distant parts of the Universe, so the wavelength of the radiation is redshifted by the expansion of the intervening space. Looking back to the time when the CMB radiation and matter were last in equilibrium with each other entails a redshift of about 1000. In other words, the wavelength corresponding to the mean photon energy of the CMB radiation today (about 2 mm) was 1000 times shorter (i.e. it was about 2 μm) when the radiation was emitted. The wavelength of the radiation has been 'stretched out' by the expansion of the Universe. A lengthening of the wavelength from 2 μm to 2 mm corresponds to a reduction in photon energy from 1 eV to 0.001 eV. So the expansion of the Universe has shifted the mean photon energy of the radiation from the infrared part of the spectrum down into the microwave region where it is observed today.

The CMB radiation is therefore a relic of the time when radiation and matter in the Universe existed in equilibrium, at a temperature of around 3000 K. The CMB photons pervade the entire Universe, wherever astronomers look. This indicates that the entire Universe was once at a much higher temperature.

A comparison between conditions in the Universe when the CMB radiation was produced and conditions today is given in Table 9.1.

Table 9.1 A comparison between conditions in the Universe when the CMB radiation was produced and conditions today.

	Conditions in the Universe when the CMB was produced	Conditions in the Universe today
temperature of the Universe	3000 K	3 K
mean wavelength of radiation	2 μm	2 mm
mean photon energy	1 eV	0.001 eV
spectral region	infrared	microwave
redshift relative to today	1000	0

9.3 The hot big bang model

The expansion of the Universe (as expressed by the Hubble relationship) tells us that objects in the Universe were closer together in the past than they are today. In other words, the Universe was *denser* in the past than it is now. The cosmic microwave background radiation tells us that the Universe was *hotter* in the past than it is now. These are the two main pieces of evidence pointing to the fact that our Universe originated in what has become known as a **big bang**.

In fact, the idea of a big bang actually pre-dates not only the discovery of the cosmic microwave background, but also Hubble's discovery of the expansion of space. In 1927, the Belgian priest Georges Lemaître had already proposed that the Universe began with an 'explosion of a primeval atom'. The idea was subsequently developed by the Russian-American physicist George Gamow who predicted the existence of the cosmic microwave background radiation, long before Penzias and Wilson discovered it. However, the name 'big bang' was not coined until 1949 when the British astronomer Fred Hoyle used the term in a BBC radio programme in order to denigrate the theory, which he disliked and which was a rival to his own 'steady-state' model of the Universe.

The now standard model for the origin and early evolution of the Universe is sometimes known as the hot big bang model. Space and time were created in this event, and space has expanded as time has progressed ever since. The story of the evolution of the Universe from the time of the big bang to the present day will be presented in Chapter 15. However, in order to discuss such an immense topic, it is necessary to appreciate the role that four fundamental interactions each play in the evolution of the Universe. In Part III, you will learn about these four interactions in turn, and you will discover how attempts are being made to unify these into a single, coherent, theory of everything.

9.4 Summary of Chapter 9

The radiation emitted by an object in thermal equilibrium has a black-body spectrum, and this continuous spectrum has a characteristic shape. The mean photon energy of a black-body spectrum is determined solely by the temperature of the body – higher temperature objects have spectra with a higher mean photon energy. A temperature of 3000 K corresponds roughly to a mean photon energy of 1 eV.

Wherever astronomers look in the Universe they see a 'glow' of radiation known as the cosmic microwave background. The black-body spectrum of this radiation corresponds to a temperature of about 3 K. This radiation has been redshifted to longer wavelengths (cooler temperatures) by the expansion of the Universe. The expansion and cooling of the Universe point to the fact that the Universe was both denser and hotter in the past. It is believed that time and space were created in an event referred to as the big bang.

PART III – WHAT RULES DOES THE UNIVERSE FOLLOW?

> In this house, we OBEY the laws of thermodynamics!
>
> Homer Simpson, cartoon character

In Chapter 15, you will get a fuller picture of what the Universe was (most probably) like when the cosmic background radiation interacted readily with matter (long before the formation of the first galaxies and stars). It is even possible to give a reasonable account of much earlier epochs in which the atoms, nuclei, leptons and quarks came into existence. But first, it is necessary to summarise what is known about the interactions of matter and radiation, from experiments performed at energies achievable on Earth. These in turn suggest some things that might occur at energies higher than can be achieved on Earth, but which were probably very common in the early Universe.

So you need to continue your investigations into the smallest structures of the Universe. This reflects the dramatic interplay between cosmology and particle physics in recent decades: each informs the other. Cosmologists have learned things from the cosmic microwave background radiation that challenge particle physicists to conjecture about interactions of particles at energies higher than those achievable in the laboratory. Particle physicists extrapolate their understanding of theories developed to explain laboratory results and then turn back to the Universe for evidence against which to test these theories.

■ Which two observed features of the Universe, described in Chapters 8 and 9, suggest that in its past it would have been a good laboratory for particle physics?

□ The Universe is expanding (Chapter 8) and cooling (Chapter 9), so in the past it was denser and hotter, and particle interactions would have been more frequent and more violent.

In Chapter 15, you will see that information from particle physics is needed to tell the best current version of the history of our Universe. It turns out that what is needed is a quantum theory of the interactions – or forces – between particles, and an understanding of how they change their character when the participating particles interact at high kinetic energies. In outlining this, four **fundamental interactions** will be referred to, as follows:

1. **Electromagnetic interactions**: these are responsible for the forces between electrons and protons in atoms, and for the emission and absorption of electromagnetic radiation, such as light. A residual electromagnetic interaction allows atoms to bind together to make molecules and so is responsible for all chemical, and ultimately biological, processes.
2. **Strong interactions**: these provide the (very) strong forces between quarks *inside* protons and neutrons. A small residual effect of the strong interactions between quarks allows protons and neutrons to bind together in the nuclei of atoms.

3 **Weak interactions**: these are responsible for processes such as radioactive beta-decay which involve both quarks and leptons, and in which particles transform from one type to another.

4 **Gravitational interactions**: these make apples fall to Earth, maintain planets in their orbits around stars, and control the expansion rate of the Universe. However, they are negligible within the atom. But when matter aggregates into huge (and electrically neutral) lumps, such as planets and stars, gravity dominates.

These four interactions are the subjects of Chapters 10–13. You will look at the way in which these interactions operate, and also at how their strengths compare. In Chapter 14, you will learn that these interactions are not as distinct as had been previously supposed. Some (perhaps all) of them may be different aspects of a more unified description of nature, appearing to us as different as ice, water, and steam, yet deriving from the same basic principles.

Remember, the aim is to understand how the Universe works. To do this you need to know about the different stages involved in building up nuclei, atoms, planets, stars and galaxies out of the material that emerged from the big bang. The key to all these processes is an understanding of the four fundamental interactions that govern the behaviour of the contents of the Universe. It is in this third part of the book, therefore, that the second big question posed in Chapter 1 is addressed – what rules does the Universe follow?

Chapter 10 Electromagnetic interactions

> There are forces in nature called Love and Hate. The force of Love causes elements to be attracted to each other and to be built up into some particular form or person, and the force of Hate causes the decomposition of things.
>
> Empedocles, Greek philosopher (*c*. 490 BC – *c*. 430 BC)

The behaviour of particles within atoms and molecules involves the interactions between charged particles, such as those which occur between electrons and protons in an atom. These electromagnetic interactions, which are at the origin of all atomic and molecular activity, have two aspects: electric forces between charged objects, and magnetic forces between moving charges (i.e. electric currents). Electromagnetic radiation (such as light) is emitted or absorbed (as photons) in processes involving these forces. These three features of electromagnetic interactions – electric forces, magnetic forces, and electromagnetic radiation – will feature in the rest of this chapter. The aim here is to understand their unification, in the modern theory of quantum electrodynamics (QED).

10.1 Electric and magnetic forces

The law describing the force of electrical attraction and repulsion was discovered by Charles Augustin de Coulomb in 1785 and can be expressed as follows:

> **Coulomb's law**: Two particles of unlike (or like) charge, at rest, attract (or repel) each other with an electric force that is proportional to the product of their electric charges divided by the square of their separation.
>
> Using Q_1 and Q_2 to represent the values of the two charges and r to represent their separation, the electric force between them, F_e, may be written:
>
> $$F_e = -k_e \frac{Q_1 Q_2}{r^2} \qquad (10.1)$$
>
> where k_e is a constant of proportionality whose value is $k_e = 9.0 \times 10^9\ \mathrm{N\ m^2\ C^{-2}}$.

Coulomb's law is therefore another example of an inverse square law, just like the relationship between brightness and luminosity introduced in Section 8.1.

Notice the minus sign in Equation 10.1. This tells that that if one particle has a positive charge and the other has a negative charge (i.e. unlike charges), there is a positive force of attraction, since the product $Q_1 \times Q_2$ is negative. Conversely, if the particles have charges with the same sign (either both positive or both negative), an overall minus sign remains, and the force between them is repulsive. In other words, unlike charges attract, and like charges repel each other.

You will now explore further what Coulomb's law implies about the forces between charged particles.

- ■ If the charge of one particle is increased by a factor of three, what happens to the force between the two charged particles?
- □ The force also increases by a factor of three, since it is proportional to the product of the two charges.
- ■ If two charged particles are moved twice as far apart, what happens to the force between them?
- □ The force is reduced by a factor of four, since it depends on dividing by the square of the separation.

Coulomb's law may also be used to investigate the strength of attraction between particles such as protons and electrons. The result of such a calculation is remarkable. The electric force between a proton and an electron at the typical separation that applies within a hydrogen atom is equivalent to about one-hundredth of the weight of a grain of sand. Yet a grain of sand contains around one hundred billion billion atoms. Now, the weight of an object is a measure of the gravitational force of attraction between the Earth and the object in question. So the electric force between a single proton and electron is about the same size as the gravitational force between the Earth and a billion billion atoms! From this, you can see that the gravitational force exerted by the Earth on a *single* atom is utterly negligible compared with the electric forces within it.

Question 10.1

Consider a nucleus of an atom of oxygen-16 and a nucleus of an atom of nitrogen-16 that are stationary and situated 1 m apart in empty space.

(a) Explain whether the electromagnetic force between the nuclei is attractive or repulsive.

(b) If the nitrogen-16 nucleus is replaced by a second oxygen-16 nucleus, explain how the strength of the electromagnetic force would change (if at all).

(c) State a way in which you could reduce the strength of the electromagnetic force to zero at a separation of 1 m, in this example.

Magnetism may be most familiar to you from the interaction of magnetic materials, such as a bar magnet and a compass needle. However, the magnetic properties of metals derive from the motion of the electrons that they contain. Nowadays, perhaps the commonest source of 'moving electrons' is any electrical apparatus, because an electric current is simply that: a flow of electrons.

Around 1820, Hans Christian Oersted observed that an electric current flowing in a wire affects a compass needle placed close to it, so demonstrating that the movement of charge produces a magnetic force, registered by the magnetic material in the compass needle. The converse effect is that a bar magnet exerts a magnetic force on moving charges. Around the same time, André-Marie Ampère showed that if two wires close to each other carry currents that flow in the same direction, the wires attract each other with a magnetic force; whereas if the currents flow in opposite directions, there is a magnetic repulsion.

10.2 Electromagnetic fields and radiation

In 1864, at a meeting of the Royal Society, James Clerk Maxwell (Figure 10.1) presented a set of equations that unified the laws of magnetism with those of electricity. His subsequent book outlining the theory, *A Treatise on Electricity and Magnetism*, published in 1873, explained all of the then known effects of electromagnetism, with the distinctive prediction that light is a form of **electromagnetic radiation**.

Figure 10.1 James Clerk Maxwell (1831–1879), was born in Edinburgh and was one of the greatest theoretical physicists of the 19th century. His book *A Treatise on Electricity and Magnetism* explained that light is an 'electromagnetic wave'. Maxwell also suggested that such waves existed beyond the infrared and ultraviolet regions of the spectrum, but unfortunately he did not live to see his prediction verified in 1888 when Heinrich Hertz produced radio waves.

The key concept to Maxwell was that of an electromagnetic *field*. Coulomb's law had described the electric force exerted by a charge Q_1 on a charge Q_2 (and vice versa) when positioned a certain distance apart. Now suppose that Q_2 is removed and replaced at the same location by a charge that is ten times larger. How does the force on this charge compare with that which was exerted on the original charge? Equation 10.1 shows that the force has the same direction, but is ten times larger. Maxwell's perspective was that a charge produces an **electric field**, spread out over the whole of space, *whether or not* there happens to be another charged particle responding to it at any particular point in space, at any particular time. The force becomes apparent only when the second charge is placed in the field of the first. Figure 10.2 gives a picture of the electric field due to a positive or negative charge. The electric field lines are directed radially outwards (or inwards), and the strength of the field is greater where the lines are closer together.

The electric force F_e that acts on a charge placed in an electric field is simply the strength of the electric field due to the first charge at that point in space, $\mathscr{E}_1$, multiplied by the value of the second charge, Q_2:

$$F_e = Q_2\,\mathscr{E}_1 \qquad (10.2a)$$

where the symbol for electric field strength $\mathscr{E}$ is called 'curly-E' to distinguish it from E used for energy. Electric field strength can be expressed in the SI unit of newtons per coulomb, N C^{-1}. Notice that the second charge introduced into the field of the first will also produce its own electric field.

■ What is the force acting on the first charge due to the electric field of the second?

□ The force is $F_e = Q_1\,\mathscr{E}_2$, and this will be equal to that in Equation 10.2a, but act in the opposite direction.

In general therefore, Equation 10.2a may be re-written in a more general form to express the electric force F_e experienced by a particle of charge Q when placed in an electric field of strength $\mathscr{E}$ as:

$$F_e = Q\,\mathscr{E} \qquad (10.2b)$$

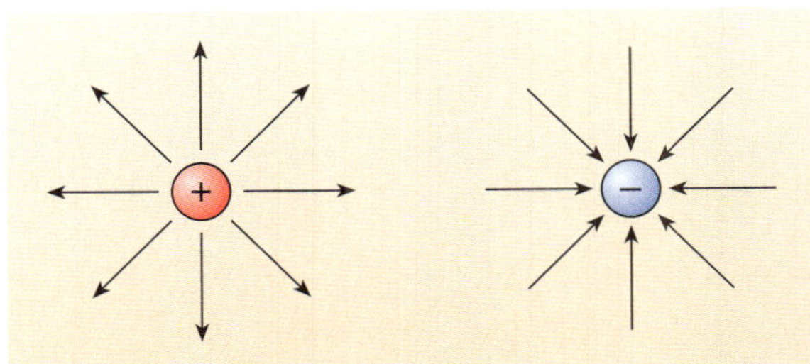

Figure 10.2 (a) The electric field of a positive charge. The field lines radiate away from a positive charge in all directions. (b) The electric field of a negative charge. The field lines converge on a negative charge from all directions.

Question 10.2

(a) What is the size of the force exerted on a charge of $Q = +2e$ when it is placed in an electric field of strength $\mathscr{E} = 200$ N C^{-1}?

(b) A charged particle experiences a force of $F_e = 3.2 \times 10^{-12}$ N when it is placed in an electric field of strength $\mathscr{E} = 5.0 \times 10^{6}$ N C^{-1}. What is the value of the charge of the particle?

(You may use the fact that the charge of a proton, $e = 1.6 \times 10^{-19}$ C.)

You may find the field idea to be an unnecessary complication, if all that results is Coulomb's original law of force. However, **magnetic fields** are much easier to work with than magnetic forces. Figure 10.3 illustrates the magnetic fields due to a straight wire and a coil of wire, each carrying an electric current. Note that, unlike electric field lines, magnetic field lines are always 'closed' loops.

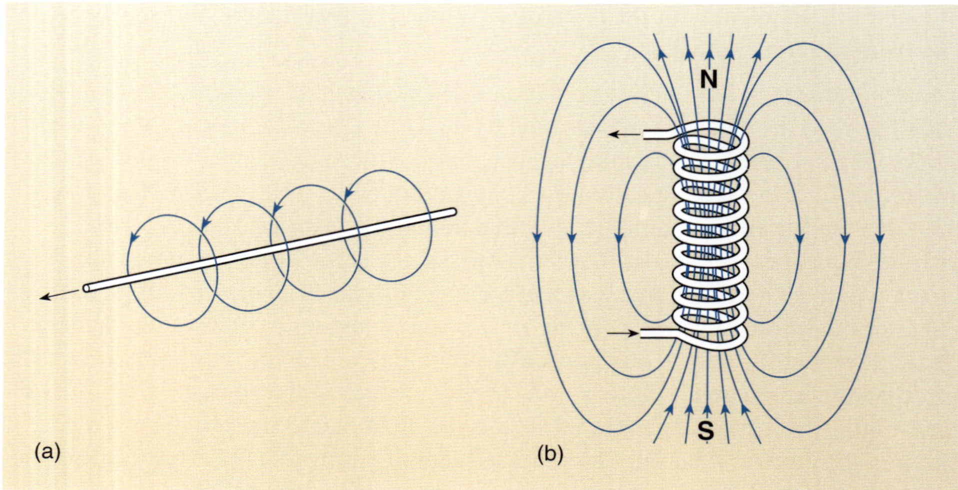

Figure 10.3 (a) The magnetic field due to a straight wire carrying a current consists of closed field lines which encircle the wire. (b) The magnetic field due to a coil of wire carrying a current consists of closed field lines which emanate from one end of the coil (the north pole, N) and loop round to the other end of the coil (the south pole, S), before passing through the centre of the coil and returning to their starting point.

So, Maxwell's concept of electric and magnetic fields was initially useful for two reasons. It meant that:

- Coulomb's law could be reinterpreted as implying that a *stationary* electric charge gives rise to an *electric* field.
- Oersted's and Ampère's discoveries could be reinterpreted as implying that a *moving* electric charge (i.e. an electric current) gives rise to a *magnetic* field.

The next discovery in this area had been made around 1830, independently by Michael Faraday and Joseph Henry. They had observed that a *changing* magnetic field produces an electric field. This phenomenon is called electromagnetic induction and is the basis of a dynamo, the principle of which is illustrated in Figure 10.4.

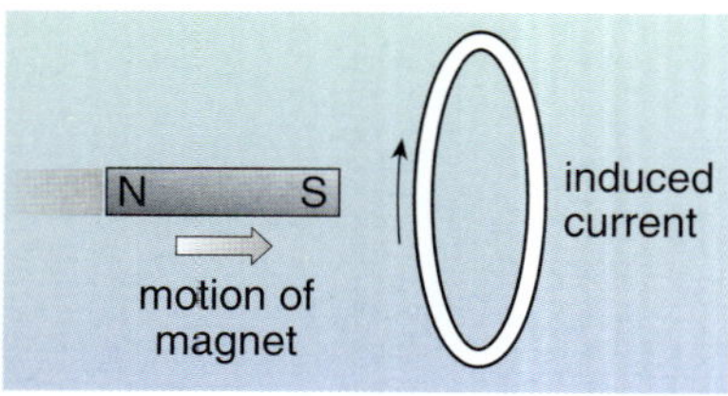

Figure 10.4 An example of electromagnetic induction. When a magnet is brought swiftly towards a loop of wire, a current is produced in the loop. The explanation is that the changing magnetic field, caused by moving the bar magnet, acts on the wire and sets up an electric field that causes electrons to move round the wire.

Maxwell's great contribution was to predict one more phenomenon: that a *changing* electric field produces a magnetic field. The previous three parts of electromagnetic theory had been developed in response to experiments: the electric field of a stationary charge came from observations by Coulomb; the magnetic field of a steady current from studies by Oersted and Ampère; and the electric field produced by a changing magnetic field from measurements by Faraday and Henry. There were no data requiring Maxwell's fourth idea, that a magnetic field can be produced by a changing electric field. Instead the prediction was motivated by a study of the equations describing the previous three effects, and by a sense that the addition made a more 'elegant' set of equations.

Yet mere elegance is (as Einstein remarked) for tailors; scientists must make predictions, or illuminate old problems, if their contributions are to endure. Maxwell's prediction was dramatic in the extreme: according to his equations, an *electromagnetic wave* could be set up, with electric and magnetic fields at right angles to each other and to the direction in which energy is transported, as shown in Figure 10.5. Furthermore, these waves would all travel through a vacuum (i.e. empty space) with the universal (constant) speed of 3.0×10^8 m s^{-1}, which is also the speed of light.

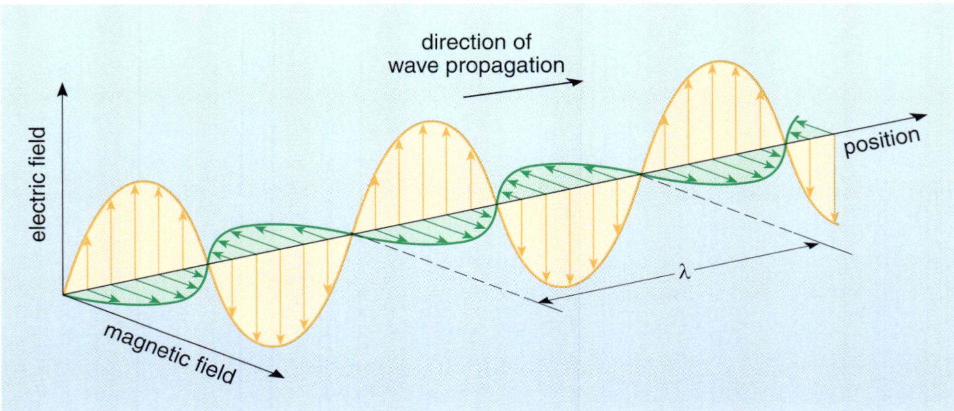

Figure 10.5 The electric and magnetic fields in an electromagnetic wave.

In other words, Maxwell had not only thrown new light on light itself, but suggested the existence of electromagnetic radiation of *any* wavelength. All the various types of radiation from radio waves and microwaves, through infrared, visible and ultraviolet radiation, to X-rays and gamma rays are merely different manifestations of the same oscillating electric and magnetic fields described by Maxwell. It is hard to imagine more far-reaching consequences of trying to tidy up a set of equations. Most of the technology that is vital to modern society depends on the understanding of electricity, magnetism and electromagnetic radiation, and all of these devices have been enabled by the insight that Maxwell provided.

Question 10.3

Summarise what was known about electric and magnetic fields before James Clerk Maxwell's work. Then summarise the key predictions made by Maxwell.

Activity 10.1 The electromagnetic world

We expect this activity will take you approximately 10 minutes.

Think back over the last 24 hours and make a list of the various items you have used that rely on using electric and magnetic forces and fields and on electromagnetic radiation. Then reflect for a moment on the importance of Maxwell's work that enabled virtually all of the technology on which modern human society depends.

Now look at the comments on this activity at the end of this book.

10.3 Quantum electrodynamics

Maxwell completed his theory of electromagnetism in 1873, and you may be wondering whether that was the last word on this phenomenon. In fact, it was not by a long way. The quantum physics of atoms is inextricably linked with the emission and absorption of electromagnetic radiation (photons). So clearly there was a need to unite Maxwell's theory of electromagnetism with the quantum model of the atom if a correct description of atoms and radiation was to be obtained. Another important development in physics in the early part of the 20th century was Einstein's theory of **special relativity**, published in 1905. A key result of this theory is that the kinetic energies of particles travelling at a substantial fraction of the speed of light do not obey the conventional formula that applies in everyday situations, and which you met in Book 3, Section 3.1. Electrons in atoms have a range of possible speeds, and the most probable speed for the electron in a hydrogen atom is around 1% of the speed of light. Even at this speed, the conventional formula for kinetic energy is in error by 0.005%. At higher speeds (which are possible in atoms), even larger errors occur if special relativity is not taken into account.

A fully consistent explanation of the properties of atoms, electrons and radiation must therefore combine *electromagnetism* with *quantum physics* and *special relativity*, to produce what is called a relativistic quantum theory of these properties. The first stage in this process was completed by Paul Dirac in 1928, the year after the famous Solvay conference photograph (Figure 3.2) was taken. To achieve a high precision in describing tiny corrections to the quantum model of the atom as a result of special relativity, Dirac predicted the existence of a new particle, with the same mass as the electron, but with positive charge. As you saw in Section 6.3, this particle is called the positron, or antielectron, and given the symbol e^+. Now, as you've also seen in Section 2.2, in the process of pair creation, high-energy photons can create electron–positron pairs (Equation 2.2b), thereby producing new particles from purely electromagnetic energy. The mass energy of an electron or positron is about 510 keV so to create an electron–positron pair requires just over 1 MeV of energy.

■ What are the typical separations between energy levels in hydrogen atoms? What is the typical energy of a photon emitted or absorbed by hydrogen atoms?

□ Energy levels in hydrogen atoms are typically separated by only a few eV. Consequently, the photons emitted or absorbed by hydrogen atoms also have energies of only a few eV.

Pair creation, therefore, requires photon energies that are about a million times greater than the energies of a few eV that are involved in the energy levels of the hydrogen atom or photons of visible light! So, at first sight, there is no need to consider positrons in atomic physics. After all, the idea of energy conservation is basic to physics. Why consider the creation of positrons, when none ever emerge from atoms, apart from the comparatively rare unstable nuclei that undergo beta-decay?

It turns out that there *is* a need to consider positrons. The conservation of energy is something that has to work out only on long timescales; on much shorter timescales energy accounting may be 'relaxed', provided the accounts are settled in the long run. The basic rule for this had been stated by Werner Heisenberg in the 1920s: a failure of energy conservation may be tolerated for a short time provided that the energy deficit multiplied by the time interval is less than the value of the Planck constant, h, divided by 4π:

$$\Delta E \times \Delta t < \frac{h}{4\pi} \tag{10.3}$$

Since the Planck constant has a value $h = 4.1 \times 10^{-15}$ eV s, the right-hand side of Equation 10.3 has a value of 3.3×10^{-16} eV s (to 2 significant figures). After the time limit specified by Equation 10.3, the energy debt must be made good. In fact, this is just another form of the Heisenberg uncertainty principle that you met in Section 4.3 in relation to how precisely positions and velocities of electrons in atoms can be known. This **energy–time uncertainty principle** embodies an important feature of the quantum world: whatever is allowed to happen will do so, sooner or later.

Try thinking of the uncertainty principle as an arrangement with your bank manager, who says you may go overdrawn by £100 for no more than 100 days, or by £1000 for no more than 10 days, or by £10 000 for a single day. In this scenario, anything seems to be allowed, provided the product of your debt and its duration is no more than 10^4 pound-days.

■ In this analogy, for what length of time could you borrow a million pounds?

□ The time is given by:

$$\frac{10^4 \text{ pound-days}}{10^6 \text{ pounds}} = 0.01 \text{ of a day, or about a quarter of an hour.}$$

■ How much could you borrow for a thousand years?

□ A thousand years corresponds to 365 000 days (to 3 significant figures), so the amount you could borrow is given by:

$$\frac{10^4 \text{ pound-days}}{365\,000 \text{ days}} = 0.027 \text{ pounds or less than 3 pence!}$$

In physics, the general feature of the energy–time uncertainty principle is that you may push it to the limit; any failure of the conservation of energy may be tolerated for a certain time, provided that the limit above holds true. The terms are not generous, but they are totally flexible, within this credit limit.

Question 10.4

To create an electron–positron pair takes about 1.0 MeV of energy.

(a) Suppose the pair was created out of *nothing*, what is the maximum time for which this energy may be 'borrowed'?

(b) How could the energy debt be 'repaid'?

The energy–time uncertainty principle led to the idea that empty space, even inside an atom, is not really empty. In informal terms, it is as if electrons and positrons were constantly appearing out of nothing, and then disappearing before their credit has run out. The space inside a hydrogen atom (and inside any other atom) is filled with transient electron–positron pairs. The (negatively charged) electrons that are created are drawn towards the (positively charged) proton at the centre of the hydrogen atom, effectively screening the charge of the proton.

If you were to measure the electric force of attraction on an electron produced by the nucleus of a hydrogen atom (i.e. a proton) when situated relatively far away from it, you would get an answer that agrees with Coulomb's law (Equation 10.1). This equation describes the effective strength of electromagnetic interactions when screening due to transient electron–positron pairs *is included*. However, if you were to measure the same effect when much closer to the nucleus, some of the transient electron–positron pairs would be further away. Consequently, there would be *less* screening, so the effective charge of the proton would appear slightly larger, and the electric force would consequently increase. In other words, at small distances, the electromagnetic interaction will appear *stronger* than it does at larger distances.

These quantum effects modify the electric force in a hydrogen atom only over distances that are about 1% of the typical separation of an electron and proton. The result is to modify the energy levels by a fractional amount that is only about 1 part in a million. Yet the discovery and explanation of such tiny effects led to the development of a whole new theory, called **quantum electrodynamics** (QED). Quantum electrodynamics is the most complete theory of electric and magnetic interactions that we possess. It was developed in the 1940s by Richard Feynman, Julian Schwinger and Sin-Itiro Tomonaga, for which they received the Nobel Prize for Physics in 1965. QED incorporates descriptions of the emission and absorption of photons, and is needed to understand many features of the subatomic world. In this theory, all electric and magnetic forces are envisaged as arising from the *exchange* of photons between charged particles. Electricity, magnetism, electromagnetic radiation and the behaviour of electrons in atoms are merely different aspects of the same phenomenon. Many confirmations of this theory have now been obtained. A schematic diagram outlining the progression towards this theory is shown in Figure 10.6.

The final surprise of QED has already been hinted at above. It concerns how the strength of the electromagnetic interaction varies depending on the energy at which the effect is investigated. When atoms are probed in high-energy experiments, distances close to the nucleus are investigated. In these regions, the strength of electromagnetic interactions is greater than the strength predicted by Coulomb's law, because shielding of the nucleus by transient electron–positron pairs is less effective. In other words, electromagnetic interactions appear *stronger* when investigated at high energies. In Chapter 15, you will see that the varying strength of fundamental interactions with energy is a crucial feature that enables us to understand the conditions that prevailed in the early Universe.

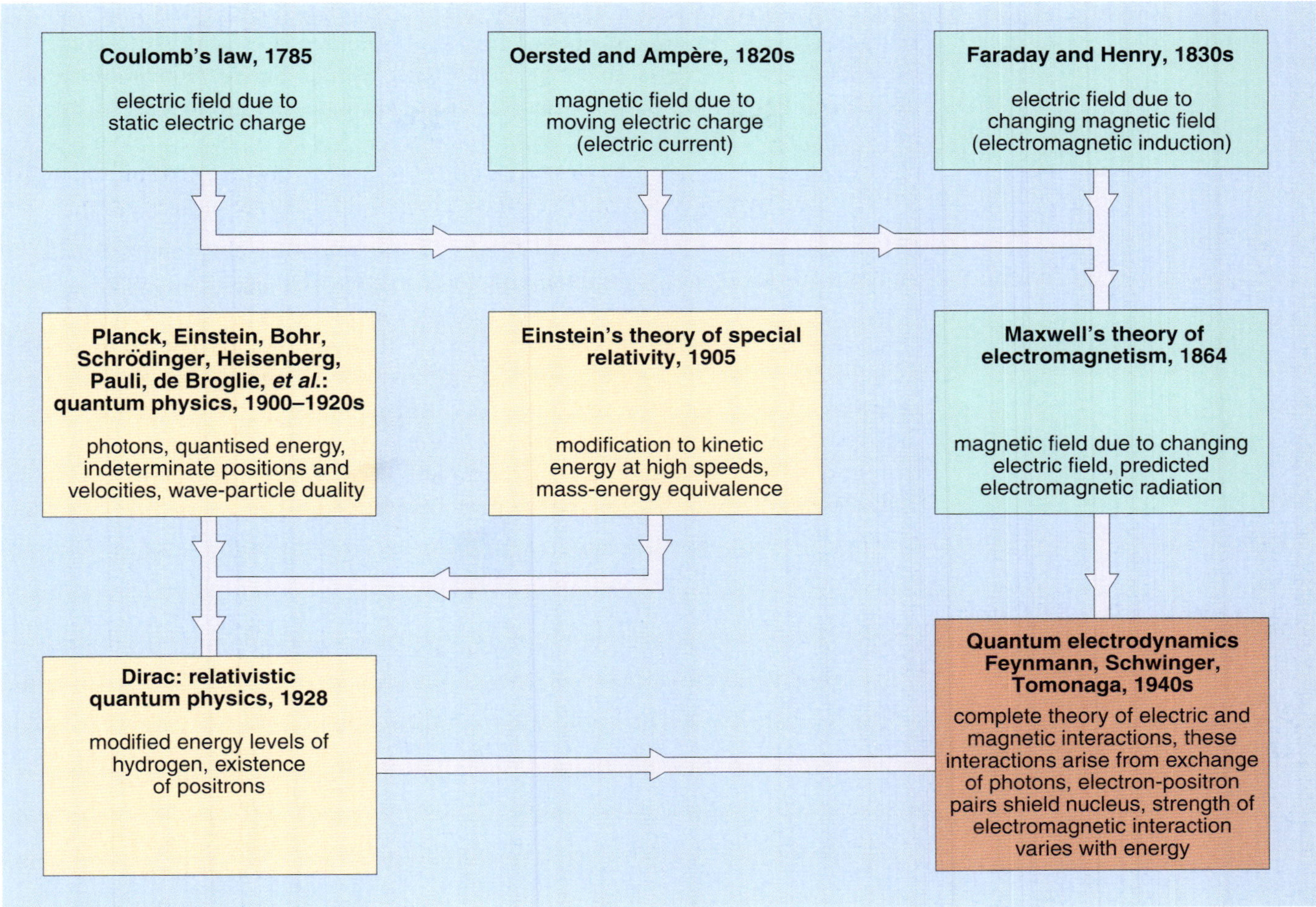

Figure 10.6 The development of the theory of quantum electrodynamics.

10.4 Summary of Chapter 10

The electric force of attraction or repulsion (F_e) between electric charges (of value Q_1 and Q_2) separated by a distance r is described by Coulomb's law. This is the force that binds electrons in atoms. Like charges repel and unlike charges attract each other. Coulomb's law may be expressed as:

$$F_e = -k_e \frac{Q_1 Q_2}{r^2}$$

where k_e is a constant of proportionality whose value is $k_e = 9.0 \times 10^9$ N m^2 C^{-2}.

Maxwell reinterpreted the work of Coulomb and of Oersted and Ampère as implying that a *stationary* electric charge gives rise to an *electric field* and that a *moving* electric charge (i.e. an electric current) gives rise to a *magnetic field*. The electric force experienced by a charge placed in an electric field is given by $F_e = Q\mathscr{E}$ where $\mathscr{E}$ is the electric field strength in units of N C^{-1}. Maxwell further combined these ideas with the discovery of Faraday and Henry that a *changing* magnetic field gives rise to an electric field and predicted that a *changing* electric field would give rise to a magnetic field. In producing his theory that unified the phenomena of electric and magnetic forces, Maxwell also predicted the existence of electromagnetic radiation that travels in empty space with speed 3.0×10^8 m s^{-1}, i.e. the speed of light.

By combining electromagnetism with special relativity and quantum physics, the theory of quantum electrodynamics (QED) was arrived at. This theory modifies Coulomb's law at very short distances, by taking account of the transient electron–positron pairs that can appear for short periods of time in empty space. These may be thought of as appearing and then rapidly disappearing, since a failure of energy conservation may be tolerated for a short time as long as the product of the energy deficit and time interval is less than the Planck constant divided by 4π, i.e. 3.3×10^{-16} eV s. QED predicts that the strength of electromagnetic interactions increases with increasing energy because at the short distances probed in high-energy experiments, the shielding effect of transient electron–positron pairs is less effective.

Chapter 11 Strong interactions

The force is strong with this one.

Darth Vader, former Jedi Knight, in the film *Star Wars*

The force that binds quarks together inside nucleons (i.e. neutrons and protons) is known as the strong interaction and has a very short range. It operates essentially only within the size of a nucleon. When two up quarks and a down quark form a proton, or when two down quarks and an up quark form a neutron, the strong interaction has, largely, done its job (in much the same way that the electric interaction between a proton and an electron does its job, so to speak, by forming a hydrogen atom). However, there is a residual strong interaction *between* nucleons, which you can imagine as 'leaking out' of the individual protons and neutrons. This is sufficient to bind them together in nuclei and is shown in Figure 11.1a. The residual strong interactions of quarks are similar in nature to the residual electromagnetic interactions between atoms that are responsible for the formation of molecules (Figure 11.1b).

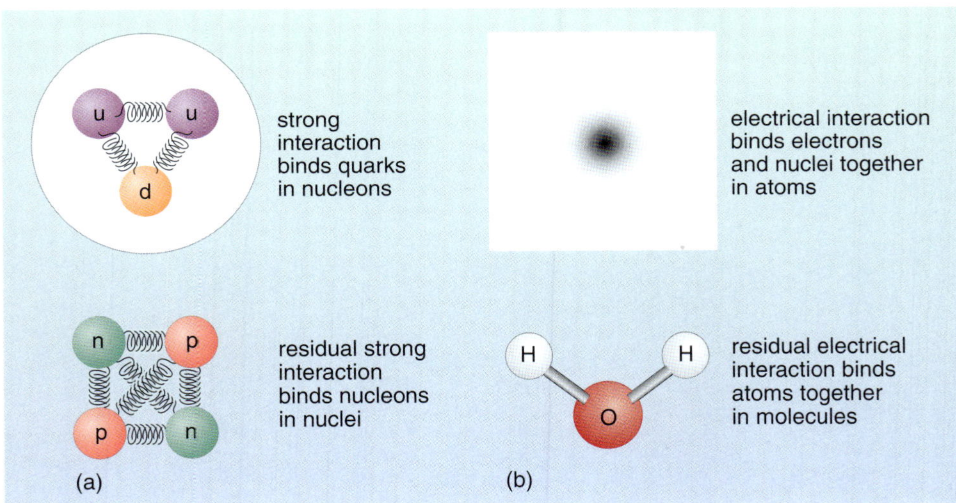

Figure 11.1 (a) The strong interaction binds quarks together in nucleons. A residual strong interaction binds nucleons in nuclei. (b) The electromagnetic interaction binds electrons and nuclei in atoms. A residual electromagnetic interaction binds atoms in molecules.

In this chapter, you will examine the first of two fundamental interactions that are likely to be the least familiar to you. The strong interaction is vital for understanding how the Universe works because it is what binds quarks together inside protons and neutrons and also what allows nuclei consisting of protons and neutrons to exist. By understanding how the interaction operates, you will be able to appreciate the vital role it played in the very early Universe. Quite simply, the strong interaction enabled the fundamental particles created out of the big bang to begin to bind together to form the more familiar nuclei of which the present-day Universe is composed.

11.1 Quarks and gluons

It is amazing just how strong the strong interaction between quarks is. At a separation of around 10^{-15} m (the typical size of a proton or neutron) the force of attraction between a pair of quarks is equivalent to the weight of a 10 tonne truck.

As might be suggested by its name, the strong force of attraction between two up quarks is much larger than the electric force of repulsion between them. It is this strong force that prevents quarks from being liberated in high-energy collisions. Free quarks are *never* seen to emerge from such processes: quarks only exist confined within baryons or mesons. It is as if they are stuck together with super glue.

Consider, for example, an experiment conducted at the Large Electron–Positron (LEP) collider at the European Organization for Nuclear Research (CERN), near Geneva, where positrons of kinetic energy 50 GeV collide with electrons with the same kinetic energy travelling in the opposite direction (Figure 11.2). In this case, an electron and a positron may annihilate each other, producing a quark–antiquark pair, with total energy 100 GeV or 10^{11} eV. These energies are certainly impressive; 10^{11} times greater than the few eV that is typical for the energy levels of the hydrogen atom and even 10^5 times greater than the combined mass-energy of the electron and positron (i.e. 1 MeV). But what can such high energies achieve, against the strong force? In fact, 100 GeV of energy can separate a quark–antiquark pair by only about 100 times the size of a proton.

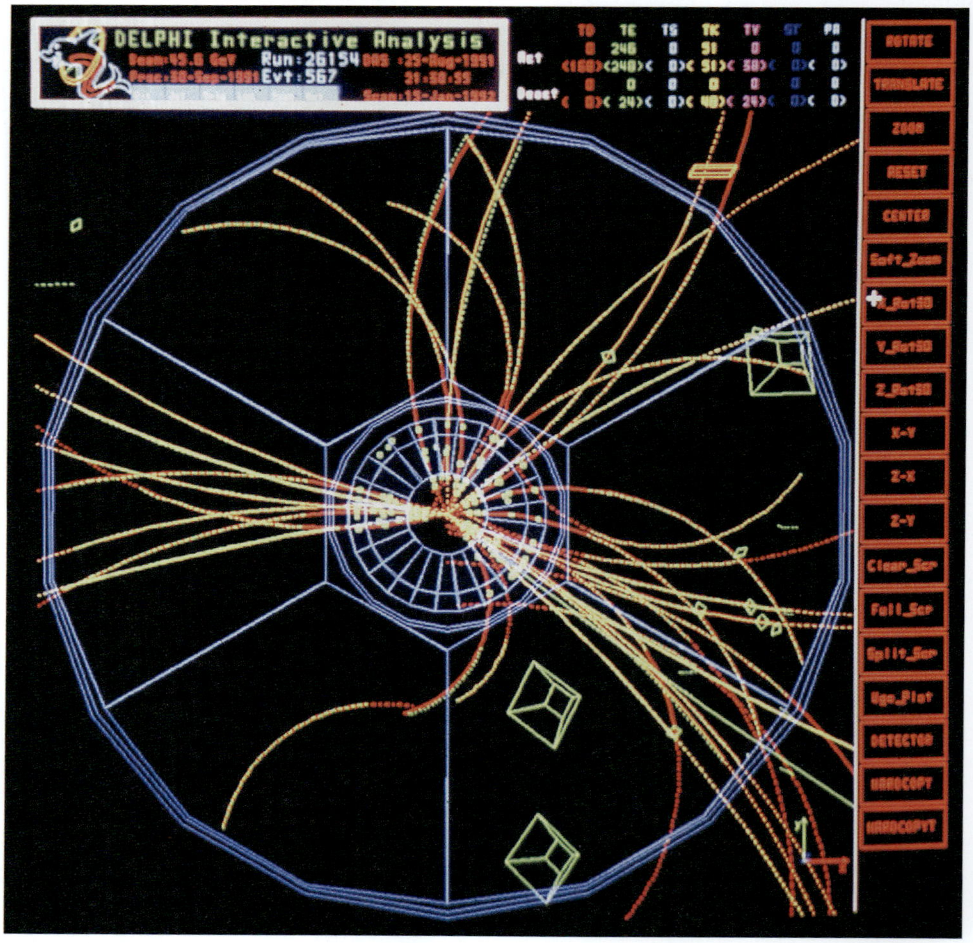

Figure 11.2 Schematic image of the particle tracks resulting from a collision in the Large Electron–Positron (LEP) collider at CERN.

But if quarks do not emerge from these collisions, what happens to the 100 GeV that is put into the collision? Energy cannot be destroyed, but can only be transformed from one kind to another (Book 3, Chapter 2). In the mess of debris that results from high-energy collisions between electrons and positrons, there is a tell-tale clue as to the original interaction that occurred at very short distances:

it often happens that hadrons emerge as a pair of jets, each jet made up of a number of hadrons. (Recall from Chapter 7 that hadrons are composite particles made of quarks and antiquarks.)

The basic interaction that produces the pair of jets is shown schematically in Figure 11.3. First, the electron (e^-) and positron (e^+) annihilate each other and can be thought of as producing what is known as a **virtual photon**. The reason for this name is that the virtual photon only has a temporary existence and immediately undergoes a pair creation event, giving rise to a quark–antiquark pair. Both the matter–antimatter annihilation event and the pair creation event are electromagnetic interactions. The virtual photon is understood merely as a way of carrying the electromagnetic force from the annihilation event to the pair creation event; it does not escape from the process and is never detectable. Similarly, the quarks and antiquarks created from the virtual photon cannot separate indefinitely. Instead, their kinetic energy and mass energy are converted into the kinetic energy and mass energy of many more matter and antimatter particles, including lots more quarks and antiquarks. Remember, as long as energy is conserved, any energy transformations are possible. The many quarks and antiquarks then combine to form a variety of hadrons, and it is only the hadrons that then emerge from the collision as a pair of jets.

It so happens that Figure 11.3 is not the whole story of what may occur in an electron–positron collision producing hadron jets; sometimes *three* jets may be produced. A schematic illustration of this situation is shown in Figure 11.4. It

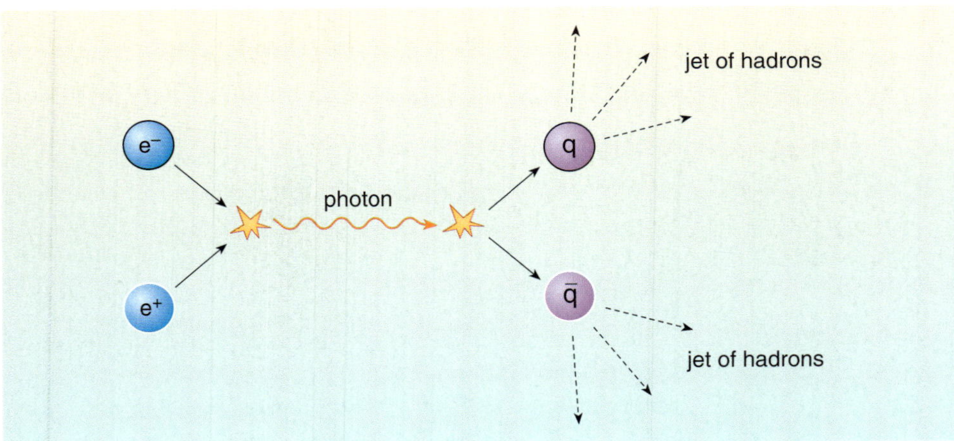

Figure 11.3 An electron and a positron mutually annihilate each other to create a virtual photon which subsequently creates a quark–antiquark pair. The energy of the quark and antiquark are then transformed into many more quarks and antiquarks, which give rise to a pair of jets of hadrons.

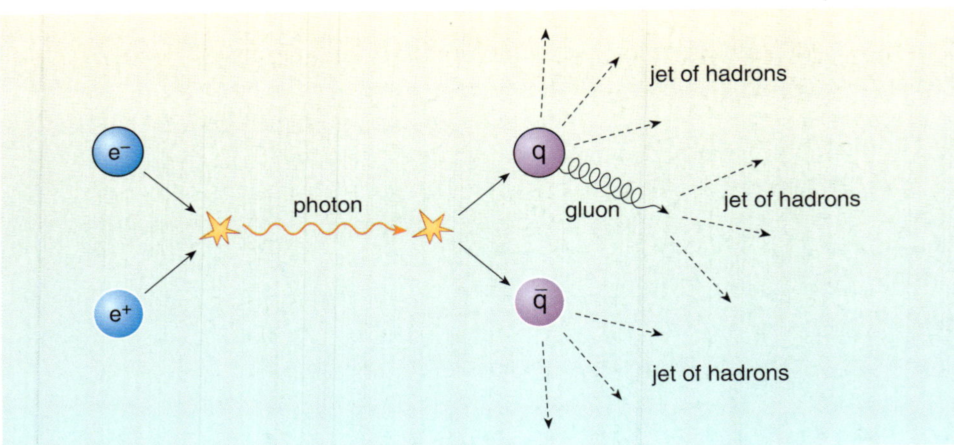

Figure 11.4 An electron and a positron mutually annihilate each other to create a virtual photon which subsequently creates a quark–antiquark pair. The quark emits a gluon, shown by the curly line. The energy of the quark, antiquark and gluon are then transformed into many more quarks and antiquarks, which give rise to three jets of hadrons.

involves a quite new ingredient, called a **gluon**. Just as photons are the quanta of energy associated with electromagnetic interactions, so gluons are the quanta of energy whose emission and absorption are regarded as the origin of strong interactions. They are responsible for 'gluing' the quarks strongly together inside hadrons. However, unlike photons, but like quarks and antiquarks, gluons cannot escape to large distances. As shown in Figure 11.4, a quark (or indeed an antiquark) may emit a gluon and in doing so the quark loses some energy. But the quark still has plenty of energy left and can go on to produce a jet of hadrons, just as already described in relation to Figure 11.3. A second jet of hadrons is produced by the antiquark, again as already described in relation to Figure 11.3. In addition, the mass energy and kinetic energy of the gluon is quickly turned into the mass energy and kinetic energy of further pairs of quarks and antiquarks. These in turn combine with each other to form various hadrons, and the hadrons produced from the gluon then give rise to a *third* jet emerging from the process.

One way of characterising the strength of strong interactions is to measure how often a three-jet outcome occurs, compared with a two-jet outcome, since a three-jet process is the one that involves a gluon. At an interaction energy of 100 GeV, there is roughly one three-jet outcome for every ten two-jet outcomes. Furthermore, it turns out that as the energy of interaction *increases*, the fraction of three-jet outcomes *decreases*. That is, there are progressively fewer outcomes in which a gluon is involved, as the energy of interaction between the colliding electron and positron is turned up.

■ What does this tell you about the strength of strong interactions?

□ The strength of strong interactions must decrease at higher energies.

So, whereas in QED the strength of electromagnetic interactions *increases* with increasing energy, the strength of strong interactions *decreases* with increasing energy.

By comparing the types of collision that involve strong interactions with those that only involve electromagnetic interactions, it is also possible to directly compare the strengths of these two interactions. At an interaction energy of 100 GeV, it has been calculated that the strength of strong interactions is about ten times stronger than the strength of electromagnetic interactions.

This may be a good point to step back for a moment and remind you *why* you are being told about the details of these processes that rely on the strong interaction. Remember, your goal is to understand how the Universe works, and in particular what happened in the very early Universe, soon after the big bang. At those very early times, the Universe was very hot and dense, so high-energy reactions, such as those just described, played a very important part in determining how the Universe evolved. The summary of strong interactions will be completed by describing the theory that explains how these interactions actually work.

11.2 Quantum chromodynamics

The quantum theory of the strong interactions between quarks and gluons is called **quantum chromodynamics** (QCD). It deals not with force directly but with the interactions between particles and quanta. In quantum electrodynamics (QED) the primary interaction is of electrically charged particles with photons. In QCD there are two basic interactions: quarks (and antiquarks) interact with gluons, and gluons also interact with themselves. It is interactions *between* gluons that are responsible for the fact that the strength of strong interactions decreases with increasing energy.

To understand why this theory is called quantum *chromo*dynamics, you should note that 'chromo' comes from the Greek word for 'colour'. The interactions between quarks and gluons are described in terms of a new property of matter that is, rather whimsically, called **colour charge**, by analogy with conventional electric charge. Just as electromagnetic interactions result from 'forces' between electrically charged particles, so strong interactions result from 'forces' between colour-charged particles. However, whereas conventional electric charge comes in only one type that can either be positive or negative, colour charge comes in *three* types, *each* of which can be 'positive' or 'negative'. These three types of colour charge are known as red, green and blue, and their opposites are antired (or the colour cyan), antigreen (or the colour magenta) and antiblue (or the colour yellow). It is important to note that colour charge has *nothing* to do with colours of light, it is merely a naming convention. By analogy with electric charge, like colour charges repel each other, and unlike colour charges attract each other.

■ To what other colour charges will a particle with a red colour charge be attracted? What other colour charges will repel it?

□ A particle with a red colour charge will be attracted to particles with green or blue colour charge, and also to particles with antired, antigreen or antiblue colour charge. It will be repelled by particles with red colour charge.

Each quark can have any one of the three colour charges, and each antiquark can have any one of the three anticolour charges. So in effect there are three versions of each type of quark: red up quarks, blue up quarks and green up quarks, for instance. (Remember, this is in addition to the conventional electric charge that quarks and antiquarks also carry.) Gluons each carry a combination of colour *and* anticolour charge (such as red–antiblue, blue–antigreen, or green–antired), although they have zero electric charge. Leptons and photons *do not* have any colour charge associated with them.

This model helps to explain many phenomena, such as why the only possible hadrons are baryons (consisting of three quarks), antibaryons (consisting of three antiquarks) and mesons (consisting of one quark and one antiquark). Each of these composite particles is *colour neutral*, that is to say it has a net colour charge of zero. Any baryon must contain one quark with a red colour charge, one quark with a green colour charge, and one quark with a blue colour charge. By analogy with conventional colours: red + green + blue = white, a neutral colour with a net colour charge of zero, as shown in Figure 11.5a.

■ What do you think are the colour charges of the three antiquarks in an antibaryon?

□ Antibaryons must contain one antiquark with an antired colour charge, one antiquark with an antigreen colour charge, and one antiquark with an antiblue colour charge. Again this gives a net colour charge of zero, as shown in Figure 11.5b.

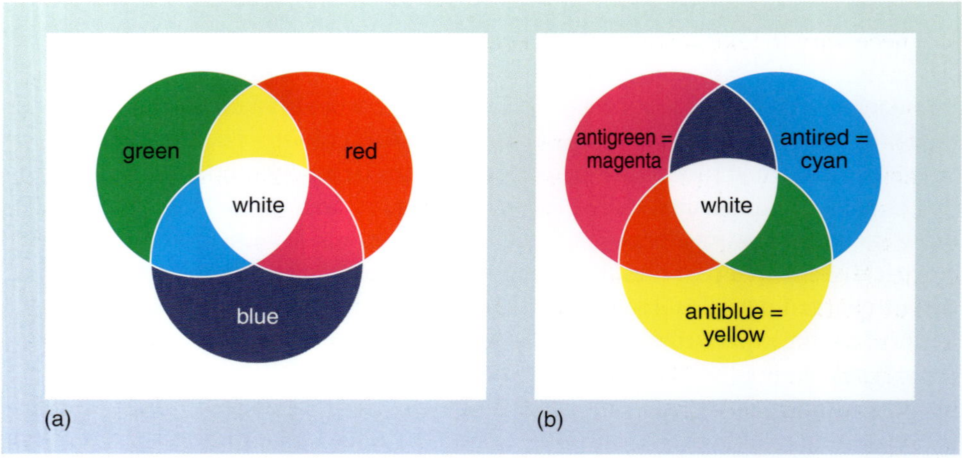

Figure 11.5 (a) Three colour charges combine to produce a net colour charge of zero (i.e. white). (b) Three anticolour charges combine to produce a net colour charge of zero (i.e. white).

Similarly, the quark–antiquark pairs that constitute a meson must have the opposite colour charge to each other: red + antired = white for instance, which is a net colour charge of zero. Only particles with a net colour charge of zero are allowed to exist in an independent state, and this explains why single quarks and antiquarks are not seen in isolation. The locking up of quarks inside hadrons is referred to as **confinement**. Gluons do not have a net colour charge of zero either, so they too do not escape from strong interactions. Instead, gluons will decay into quark–antiquark pairs, which in turn create further hadrons.

The following points serve to highlight the differences between QED and QCD:

- In QED, photons interact with electrically charged particles and their antiparticles, but *not* directly with other photons; in QCD gluons interact with quarks and antiquarks, and *also* directly with other gluons, since all of these particles possess colour charge.
- Photons and leptons escape from QED processes; gluons and quarks do *not* escape from QCD processes. Instead, they give rise to jets of hadrons (composite particles made of quarks) which do escape. The confinement of quarks inside hadrons results from the fact that only particles with a net colour charge of zero can exist in isolation.
- In QED, transient electron–positron pairs cause the effective strength of electromagnetic interactions to *increase* at the short distances that are probed in high-energy experiments; in QCD the interactions between gluons cause the strength of strong interactions to *decrease* at higher energies.

The great success of QED and QCD was to recognise that you cannot have one of these differences without the other two. The self-interaction of gluons is responsible for the weakening of the strong force at higher energies, and also for confinement of quarks and gluons. Correspondingly, the fact that photons do not interact directly with other photons is related to the increasing strength of electromagnetism at higher energies, and accounts for the fact that electrons and photons emerge from atoms, in exchange for modest amounts of energy.

So, when comparing the strengths of electromagnetic and strong interactions, it is necessary to take account of the energy scales that are involved, and the associated scales of distance, bearing in mind that higher energies probe shorter distances. As noted earlier, at an energy of 100 GeV, electromagnetic interactions are ten times weaker than strong interactions. But who knows what will happen at energies vastly higher than those achieved at the best laboratories on Earth? To make sense of the early Universe we need to know the answer. Fortunately the theories predict their own fates. If no new phenomenon intervenes, the strength of the QED interaction is condemned to increase as energy increases, while that of QCD must forever decrease as energy increases. It is therefore possible to estimate a rough value for the energy at which the two theories would have comparable strengths. The answer is thought to be about 10^{15} GeV. This is a million million times greater than collision energies currently attainable in high-energy particle physics laboratories. Yet cosmologists envisage early epochs of the evolution of the Universe when such collision energies were possible.

Question 11.1

Sketch a graph to summarise how the strengths of the electromagnetic and strong interactions vary between 100 GeV and 10^{15} GeV.

Question 11.2

Bearing in mind what you have learned about electromagnetic interactions, strong interactions, electric charge and colour charge, complete Table 11.1, which compares the properties of the fundamental constituents of the world around you, in terms of their electromagnetic and strong interactions.

Table 11.1 The properties of some fundamental particles.

Particle	Electric charge	Colour charge	Quanta with which the particle interacts
electron			
electron neutrino			
up quark			
down quark			
photon			
gluon			

11.3 Summary of Chapter 11

The strong force between quarks in hadrons is comparable to the weight of a 10 tonne truck. It confines quarks to nucleons, which are approximately 10^{-15} m in size. The forces between nucleons are a leftover effect of this strong interaction, and are of correspondingly short range. The quantum of the strong interaction in the theory of quantum chromodynamics is the gluon.

QCD is so called because it assigns to quarks a property known as colour charge. Each quark can have one of three types of colour charge (red, green or blue), and each antiquark can have one of three types of anticolour charge (antired, antigreen or antiblue). Composite particles made of quarks and antiquarks are all 'colour neutral'. The crucial difference between QCD and QED is that gluons interact directly with gluons, whereas photons do not interact directly with photons. This is explained in terms of both quarks and gluons possessing colour charge. There are two important consequences: quarks and gluons are not observed in isolation, and the strength of QCD interactions decreases as energy increases. Conversely, the effective strength of QED interactions increases as energy increases. These trends are expected to continue, with QCD and QED interactions becoming comparable at energies that are a million million times greater than can be studied at the present time in high-energy physics laboratories. Such huge energies (about 10^{15} GeV) are believed to have occurred in the very early Universe.

Chapter 12 Weak interactions

> There is a theory which states that if ever anybody discovers exactly what the Universe is for and why it is here, it will instantly disappear and be replaced by something even more bizarre and inexplicable. There is another theory which states that this has already happened.
>
> Douglas Adams, British author (1955–2001) in *The Restaurant at the End of the Universe*

Weak interactions manifest themselves as reactions, or decays, in which some particles may disappear, while others appear. There is no structure that is bound by a 'weak (nuclear) force'. Weak interactions are vital for understanding how the Universe works, as they are responsible for most of the reactions in the very early Universe by which particles changed from one sort to another. They are therefore largely responsible for the overall mix of particles from which the current Universe is made up.

The most common example of a weak interaction is beta-decay occurring in an atomic nucleus. In fact, as you saw in Section 6.3, there are three related processes, each of which is a different type of beta-decay.

■ What happens in the three types of beta-decay?

□ In a beta-minus decay, a neutron in the nucleus transforms into a proton with the emission of an electron and an electron antineutrino.

In a beta-plus decay, a proton in the nucleus transforms into a neutron with the emission of a positron (antielectron) and an electron neutrino.

In an electron capture process, a proton in the nucleus captures an electron from the inner regions of the atom and transforms into a neutron with the emission of an electron neutrino.

In each of these three processes, therefore, the nucleus involved will *change* from one type of element to another, as a result of either increasing or decreasing its proton content by one. Schematically:

beta-minus decay:

$$n \longrightarrow p + e^- + \overline{\nu}_e \qquad (12.1a)$$

beta-plus decay:

$$p \longrightarrow n + e^+ + \nu_e \qquad (12.1b)$$

electron capture:

$$p + e^- \longrightarrow n + \nu_e \qquad (12.1c)$$

12.1 The weak interaction compared

As you have seen in Chapters 10 and 11, electromagnetic interactions involve electrically charged leptons (e.g. the electron), quarks (all of which are electrically charged), and hadrons which are all made from quarks (e.g. a proton, p = uud). Strong interactions involve only particles that possess colour charge, namely quarks and gluons, as well as composite particles made from quarks. Neutrinos, which are electrically neutral leptons, are involved in neither electromagnetic interactions nor strong interactions, since they possess neither electric charge nor colour charge.

The one interaction in which neutrinos do participate is the weak interaction. The weak interactions of neutrinos from nuclear beta-decays are, as the name suggests, rather feeble. A substantial amount of the energy released inside the Sun escapes, innocuously, as the kinetic energy of neutrinos. The vast majority of them pass through the Sun without interaction. In fact, around 70 billion neutrinos from the Sun pass through each square centimetre of your body every second, without you ever noticing. However, the probability of a neutrino interacting with matter increases with the kinetic energy of the neutrino. Beams of neutrinos with kinetic energies of the order of 100 GeV are readily obtained, as decay products, at high-energy particle accelerator laboratories. At such kinetic energies, neutrinos interact as readily with a target as do electrons. So weak interactions have roughly the same strength as electromagnetic interactions at an energy of 100 GeV, and both are about 10 times weaker than strong interactions at this energy.

Table 12.1 serves to compare the various features of the strong, electromagnetic and weak interactions. The entries in the table for the first two interactions record what you learned in Chapters 11 and 10, respectively. The key feature of these is that the electromagnetic interactions of QED involve quanta called photons, while the strong interactions of QCD involve quanta called gluons. Photons interact only with particles that are electrically charged, so neutrinos are immune to them. Gluons interact only with particles that have colour charge, so all leptons are immune to them. What makes weak interactions so important is that they

Table 12.1 Comparing three interactions.

Interaction	Acts on	Particles experiencing	Quanta
strong	colour charge	all quarks (and their antiparticles); gluons	gluons
electromagnetic	electric charge	all electrically charged particles (and their antiparticles)	photons
weak	flavour	all quarks and their antiparticles; all leptons and their antiparticles; W^+, W^-, Z^0	W^+, W^-, Z^0

involve all six flavours of quark (u, d, c, s, t, b), all three electrically charged leptons (e^-, μ^-, τ^-), all three neutral leptons (ν_e, ν_μ, ν_τ), and all the corresponding six antiquarks and six antileptons, that you read about in Chapter 7. As you will see below, weak interactions enable quarks to change flavour into other quarks, and allow leptons to change flavour into other leptons. This is made possible by the quanta of the weak interaction, known as W and Z bosons, which are discussed next.

12.2 W and Z bosons

Just as photons and gluons are the quanta involved in electromagnetic and strong interactions respectively, so weak interactions involve other quanta, known as **W bosons** and **Z bosons**, as shown in Table 12.1. In fact, there are two types of W boson, one with negative electric charge, the W^- boson, and one with positive electric charge, the W^+ boson. The two (charged) W bosons each have a mass of about 80 GeV/c^2 whereas the (neutral) Z boson has a mass of about 90 GeV/c^2. In weak interactions, W and Z bosons interact with each other, as well as with all quarks and leptons. The Universe would be an impossibly boring place without them. To see what is involved, you will look at a few of the many interactions that are allowed by the theory of weak interactions.

As noted above, the beta-minus decay of a nucleus occurs when a neutron turns into a proton, with the emission of an electron and an electron antineutrino. At most, a few MeV of energy are released in this process, corresponding to the difference in mass between the original nucleus and the resultant nucleus. At the quark level, the explanation is that a down quark turns into an up quark. The process is illustrated in Figure 12.1. A down quark (d) with electric charge $-\frac{1}{3}e$ is transformed into an up quark (u) with electric charge $+\frac{2}{3}e$. A W^- boson is emitted with electric charge $-e$, thereby conserving electric charge in the process. The mass energy of the W^- boson is about 80 GeV, so it cannot possibly emerge from the nucleus as there are only a few MeV of energy available. However, it can exist for a very short time, consistent with the energy–time uncertainty principle. It subsequently decays to produce an electron (e^-) and an electron antineutrino ($\overline{\nu}_e$), setting the energy accounts straight.

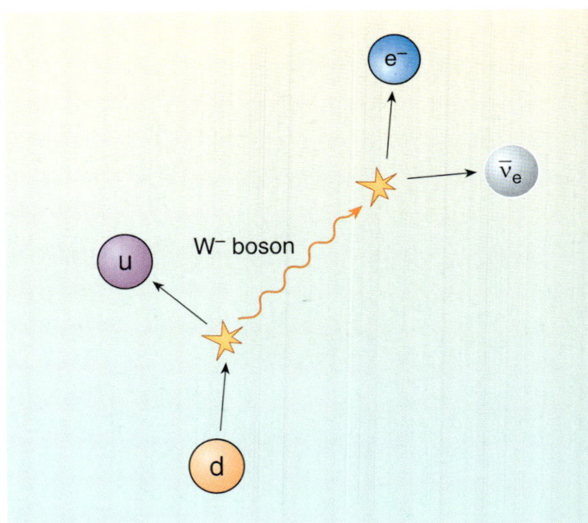

Figure 12.1 A beta-minus decay process involves the creation and disappearance of a W^- boson. A down quark decays into a W^- boson and an up quark. The W^- boson subsequently decays into an electron and an electron antineutrino.

In weak interactions, the total number of quarks minus the total number of antiquarks is the same both before and after the interaction. The number of leptons is also conserved. In the example of beta-minus decay, there are no leptons initially present, and after the interaction there is one lepton and one antilepton – a net result of zero again. This is the explanation for why neutrinos and antineutrinos are produced in beta-decays. If they were not, then the rule of lepton conservation would be violated. Notice also that the production of a charged lepton is always accompanied by the corresponding flavour of neutrino.

In *all* weak interactions:

- electric charge is conserved
- the number of quarks minus the number of antiquarks is conserved
- the number of leptons minus the number of antileptons is conserved
- flavour changing of either quarks or leptons is allowed, as long as the above three rules are obeyed.

Question 12.1

The antiparticle of a W^- boson is a W^+ boson, with the same mass, but the opposite electric charge. Following the example of beta-minus decay above, explain how beta-plus decay involves the creation and demise of a W^+ boson. Check that electric charge is conserved, that the number of quarks minus the number of antiquarks is conserved, and that the number of leptons minus the number of antileptons is also conserved.

The third type of quantum involved in weak interactions is the Z^0 boson with zero electric charge. An example of the type of reaction involving the Z^0 boson is illustrated in Figure 12.2. Here a collision between an electron (e^-) and a positron (e^+) leads to the production of a muon neutrino (ν_μ) and a muon antineutrino ($\overline{\nu}_e$). Notice that there is one lepton and one antilepton both before and after the interaction.

As noted above, the mass of a Z^0 boson is about 90 GeV/c^2, so an energy of 90 GeV is needed to create one. By selecting the energy of the electron and positron beams in the LEP collider to be 45 GeV each, so that the total energy of 90 GeV matched that required to create Z^0 bosons, a high rate was achieved for the production of neutrino–antineutrino pairs in the process shown in Figure 12.2. The experiment produced an important piece of information for understanding the early Universe: there are no more types of neutrino than the three already discovered and listed in Chapter 7, i.e. ν_e, ν_μ and ν_τ. Had there been a fourth type of neutrino, the rate of electron–positron annihilation would have been higher than that observed, by an amount significantly greater than the sensitivity of this high-precision experiment. As noted in Section 1.1, this was good news for cosmologists, who needed this information to calculate the rate at which nuclei were formed when the Universe was a few minutes old. Knowing that there are only three types of neutrino, cosmologists are able to compute the fraction of nucleons that survived as neutrons in (mainly) helium nuclei, a few minutes after the big bang.

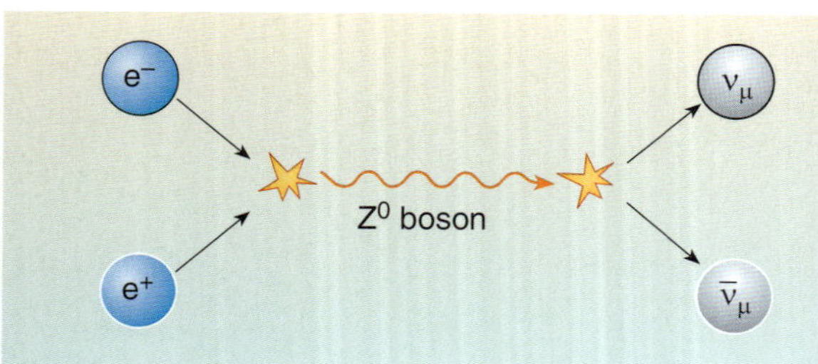

Figure 12.2 An electron–positron pair undergo annihilation, creating a Z^0 boson which subsequently decays to create a muon neutrino and muon antineutrino pair.

12.3 The survival of the neutron

Apart from hydrogen, nuclei made solely of protons cannot exist. Neutrons are necessary to make nuclei stable, so the neutron is vital to our Universe. Without it there would be only a single element, hydrogen, making chemistry extremely dull, as it would be limited to a single molecule, H_2, with no one to study it!

The rules of strong interactions allow the construction of a neutron (udd) in the same manner as a proton (uud). Indeed, as you will see in Chapter 15, in the first moments of the Universe it is believed that protons and neutrons were created in *equal* numbers. Nowadays, however, the Universe as a whole contains only about one neutron for every seven protons, and the vast majority of those neutrons are locked up inside helium nuclei. Clearly then, at some stage, neutrons have 'disappeared' from the Universe. How has this happened?

The mass energy of a free neutron is about 1.3 MeV *larger* than that of a free proton. This energy difference exceeds the mass energy of an electron (which is about 510 keV or 0.5 MeV) and means that free neutrons (i.e. neutrons not bound within atomic nuclei) can undergo beta-minus decay (as shown in Figure 12.1):

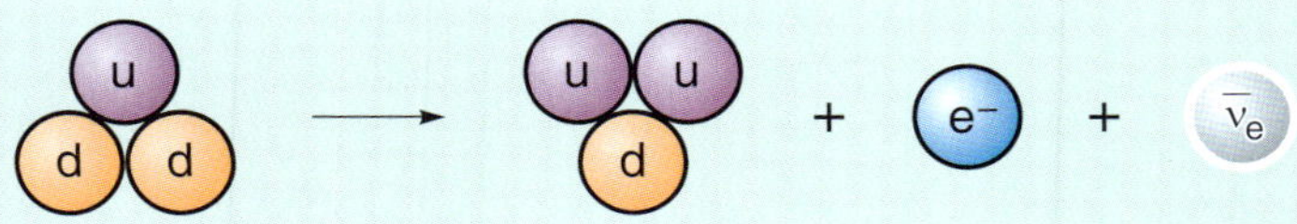

(12.2)

This is believed to be the mechanism by which the proportion of neutrons in the Universe decreased from one in every two hadrons soon after the big bang, to only around one in seven today. Once neutrons are incorporated into helium nuclei they are immune from beta-minus decay, as helium nuclei are stable.

Yet there is still a puzzle. If free neutrons can decay into protons, how did the neutrons form helium nuclei in time to avoid the fate of decay that affects them when they are free? It was indeed a question of timing. As you will see in Chapter 15, the temperature of the Universe had fallen to a value that allowed the formation of helium nuclei only a couple of minutes after the big bang. Since free neutrons survive for about 10 minutes before decaying, there were still plenty of them around at this time, and all those that had not yet decayed into protons were rapidly bound up into helium nuclei. But if free neutrons only survived for, say, one second, there would not have been many neutrons left to form nuclei a few minutes after the big bang. The vast majority of them would have long since decayed into protons. The relatively long lifetime of a free neutron is due to the fact that weak interactions (such as beta-minus decay) truly are weak, and therefore occur only rarely at low energies.

So there is a vital condition for life as we know it: weak interactions must be truly weak at low energies. If they were as strong as electromagnetic interactions at low energies, beta-minus decay processes would happen much more readily and the lifetime of a free neutron would be much shorter. As a result, the vast majority of the neutrons in the Universe would have decayed before it became possible for them to find safe havens in atomic nuclei, and there would have been no elements other than hydrogen in the Universe! Yet, at high energies, such as the 100 GeV of the LEP collider, weak interactions are comparable in strength to electromagnetic interactions, and so are only ten times weaker than strong interactions, as noted in Section 12.1. How is this trick pulled off?

It turns out to result from the large masses of the W and Z bosons, which (as mentioned earlier) are each of the order 100 GeV/c^2. In order for any weak interaction to occur, a W or Z boson must be created. But it is difficult to produce the massive W and Z bosons when the available energy is only 1 GeV. Consequently, at an energy scale of 1 GeV, where they were first investigated, weak interactions really are weak. In contrast, at an energy scale of 100 GeV, weak interactions are not so weak. At this energy, the strength of the weak interaction is roughly the same as that of electromagnetic interactions and W and Z bosons are easily created from the energy available. Going down a factor of one hundred in energy, from 100 GeV to 1 GeV, gives a huge decrease in the rates of weak processes. At an energy scale of 1 GeV, the strength of weak interactions is a hundred million times smaller than it is at 100 GeV.

Question 12.2

In a few sentences, summarise how the strengths of the three interactions that you have read about so far vary with increasing reaction energy.

12.4 Summary of Chapter 12

Weak interactions are responsible for processes, such as beta-decay, in which quarks may change flavour, and lepton–antilepton pairs may be created. They also allow leptons to change into other leptons. They involve quanta known as W bosons and Z bosons, with masses of about 80 GeV/c^2 and 90 GeV/c^2 respectively.

Weak interactions are weak only at low energies, where there is insufficient energy to create W and Z bosons easily. At an energy scale of 100 GeV, the strength of weak interactions is comparable to that of electromagnetic interactions, i.e. only about ten times less than that of strong interactions.

Free neutrons decay into protons after about 10 minutes; this is a weak interaction involving W bosons. The survival of some neutrons until helium nuclei formed in the early Universe was possible only because of the relatively long lifetime of free neutrons. This is a consequence of the weakness of weak interactions at low energies, which in turn results from the large mass of the quanta involved.

Chapter 13 Gravitational interactions

> 'Tis like this gravity, which holds the Universe together, & none knows what it is.
>
> Ralph Waldo Emerson, American writer (1803–1882)

You may think it odd that gravitational interactions are presented last on the list of the four fundamental interactions. After all, was not gravity explained satisfactorily by the work of Isaac Newton in the 17th century? In fact, there are two, rather distinct additions that need to be made to Newton's description of gravity. One, involving relativity, was completed by Albert Einstein in 1915. The other, involving quantum physics, has hardly begun.

13.1 Newton's gravity

The law that describes the gravitational force of attraction was discovered by Isaac Newton (Figure 13.1), in around 1666.

Figure 13.1 Isaac Newton (1642–1727) was probably the greatest scientist who has ever lived. During the years 1665–1666 when Trinity College, Cambridge was temporarily closed due to plague, he had his most productive and creative period working at his family home in Lincolnshire. He made fundamental advances in mathematics, essentially creating the subject of calculus; he used a glass prism to demonstrate that white light is a mixture of colours; and he began to consider the possibility that gravity might be a universal phenomenon holding the Moon in its orbit around the Earth and the Earth in its orbit around the Sun.

Newton's law of gravity: Two particles attract each other with a gravitational force that is proportional to the product of their masses divided by the square of their separation.

Using m_1 and m_2 to represent the values of the two masses and r to represent their separation, the force between them, F_g, may be written:

$$F_g = G\frac{m_1 m_2}{r^2} \qquad (13.1)$$

where G is a constant of proportionality, known as Newton's universal constant of gravitation, whose value is $G = 6.7 \times 10^{-11}$ N m^2 kg^{-2}.

Notice the close similarity between Newton's law of gravity and Coulomb's law for electrical force (Equation 10.1). Where Coulomb's law involves the product of two electric charges, Newton's law involves the product of two masses. Also, both forces decrease with the inverse of the square of the separation between the particles. Following Coulomb's law and the relationship between brightness and luminosity (Equation 8.1), Newton's law is therefore the third example of an inverse square law that you have met.

This is where the similarity between Coulomb's law and Newton's law ends. It turns out that the gravitational force of attraction between subatomic particles, such as protons and electrons is negligible. Gravitational forces are 43 orders of magnitude smaller (i.e. 10^{43} times smaller) than electric forces, between individual electrons. Recall from Chapter 1 that 45 orders of magnitude in length scales separate quarks from quasars; here there are 43 orders of magnitude separating the strength of the gravitational force from the strength of the electrical force.

Gravity comes into its own only when there are large aggregates of particles, feeling no other force. The strong and weak interactions of nuclei have very short ranges, so they make no contribution to the force between, say, an apple and the Earth. But why should gravitational forces dominate in this situation instead of electric forces?

■ Does the gravitational force act over larger distances than the electric force?

□ No, Coulomb's law and Newton's law imply that both gravitational and electric forces act over large distances with the *same* kind of inverse square law.

■ Do the electric forces cancel out somehow?

□ Yes, electric forces can be attractive *or* repulsive because objects can possess either positive or negative electric charge, and like charges repel whilst unlike charges attract. Gravitational forces are *always* attractive – there is no such thing as a repulsive gravitational force. The reason for this is that mass only comes in one form – 'negative mass' and 'antigravity' remain in the realm of science fiction.

So, the reason that gravity dominates the interaction between an apple and the Earth is that they are both electrically neutral, to very high accuracy. In order for the electric force of repulsion between an apple and the Earth to be similar to the gravitational force of attraction between them, only 1 atom in every hundred billion billion would have to lose an electron. We owe the downwards fall of the apple to the fact that matter is electrically neutral to an accuracy far better than 1 part in 10^{20}.

As a consequence of gravity, therefore, objects near to the surface of the Earth possess a property that we refer to as weight. The weight of an object is defined as the magnitude of the force of gravity acting on that object. In other words, the weight of an object of mass m_1 at the surface of the Earth is given by Equation 13.1, when m_2 is set equal to the mass of the Earth, and r equal to its radius. As you know from Book 3, Section 5.1, the weight of an object of mass m_1 is also given by:

$$F_g = m_1 g \qquad (13.2)$$

where g is the acceleration due to gravity at the surface of the Earth, which is about 9.8 m s^{-2}.

Question 13.1

Combine Equations 13.1 and 13.2 to calculate the mass of the Earth. (You may assume that the radius of the Earth is 6.4×10^6 m and $G = 6.7 \times 10^{-11}$ N m^2 kg^{-2})

The expression for the gravitational force $F_g = mg$ is rather similar to the expression for the electric force $F_e = Q\mathscr{E}$ that you met in Section 10.2. The first refers to the force acting on a particle of mass m in a region characterised by an acceleration due to gravity g; the second refers to the force acting on a particle of charge Q in a region characterised by an electric field of strength $\mathscr{E}$. The concepts are clearly very similar here; gravitational forces act on objects with mass, electrical forces act on objects with electric charge.

Question 13.2

A tiny droplet of oil with a mass of 6.4×10^{-15} kg remains stationary in mid-air between two metal plates when there is an electric field strength of 4.9×10^{4} N C^{-1} between the two plates. What is the electric charge of the oil drop? Assume that the acceleration due to gravity, $g = 9.8$ m s^{-2}. (*Hint*: since the oil drop is stationary, the net force acting on it is zero, and the gravitational force (acting downwards) must be balanced by the electrical force (acting upwards).)

13.2 Einstein's gravity

In the period between 1905 and 1915, Albert Einstein (Figure 13.2) wrestled with the consequences of a feature of gravity, which had been crystal clear since the time of Newton: you cannot use the motion of an apple, under gravity alone, to weigh the apple.

Figure 13.2 Albert Einstein (1879–1955) is widely agreed to have been the greatest physicist of the 20th century. In 1905 he published four of the most influential papers in the history of physics. Two of these outlined his special theory of relativity. Over the next few years it became clear to Einstein that an extension of his earlier work, a *general* theory of relativity, would also be a new theory of gravitation. The general theory of relativity, one of the greatest intellectual achievements of the century and a cornerstone of modern cosmology, was finally published in 1916.

As you know, *mass* is a physical property that quantifies the amount of matter in a body. In contrast, *weight* is a physical property that actually tells us the strength of the gravitational force acting on a body, expressed by Equation 13.2 for objects at the surface of the Earth. Weight therefore depends on the mass of a body *and* on, say, the planet on which the body is situated. A 100 g apple and a 50 g apple have different masses and different weights; however, on Earth the two apples fall to the ground in almost the same manner. In fact they would fall with exactly the same acceleration were it not for the (frictional) force of air resistance. In a vacuum, it is impossible to determine whether a hammer weighs more or less than a feather merely by watching them fall, since they do so with the same acceleration. The reason for this is that weight is directly proportional to mass. Doubling the mass of an object also doubles its weight, i.e. the gravitational force acting on it. The acceleration with which objects fall to the ground on Earth has the same value (about 9.8 m s^{-2}) and so tells us *nothing* about either the weight or the mass.

Einstein therefore decided that it was better to think about the regions of space and time in which this motion occurs, rather than considering forces and weights. This is not the place to describe how he went about that. It is enough to say that he was brilliantly successful, in a mission that few people had even thought about. Einstein's theory of **general relativity** reproduced all the old results of Newton, but without even using the idea of weight. The crux of general relativity is the interaction between 'space' and the 'matter' within it. One of its basic conclusions is that objects that possess mass change the geometry of space, causing it to be 'curved'. The curvature of space then controls the movement of material objects within it. It has been said that 'matter tells space how to curve' and 'space tells matter how to move'. Gravitational effects arise as a result of the curvature of space. Taking the Solar System as an example, the Sun causes space to curve in its vicinity. The orbits of the Earth and the other planets are then a consequence of their movement through this curved space.

An analogy that is often used is the idea of placing heavy ball bearings on a taut rubber sheet, as shown in Figure 13.3. The rubber sheet is a two-dimensional representation of the three-dimensional space of the real Universe. The large balls placed on the sheet represent massive objects such as the Sun, which distort the space in their vicinity. The small ball bearings represent less massive objects such as planets, which move in response to the curved space through which they travel.

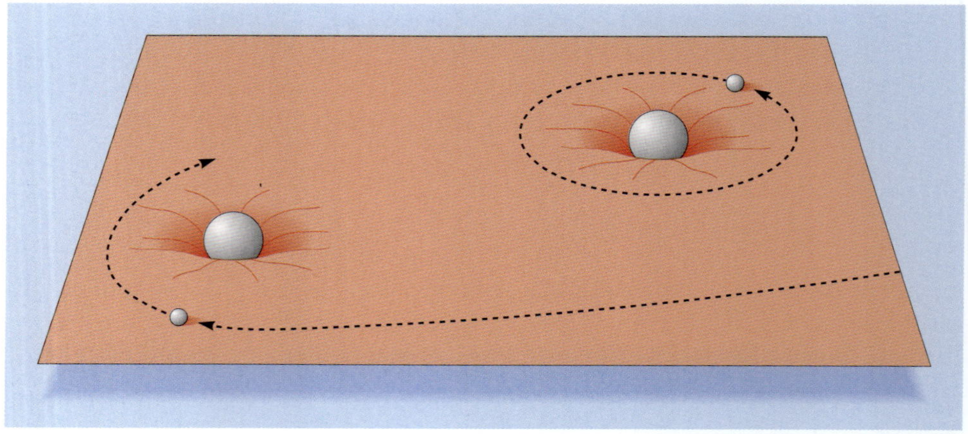

Figure 13.3 A two-dimensional analogy of the curvature of space caused by the presence of massive objects.

Does that mean that Newton's idea of massive objects producing a 'force' of attraction is wrong? Well, not exactly wrong – Newton's law of gravitation is able to predict what is actually observed in a wide variety of situations. But it is only an approximation to the real world and no longer provides satisfactory answers when the masses involved become very large. Similarly, whilst Einstein's general relativity can cope with such situations accurately, that too fails when physicists try to unite it with the ideas of quantum physics. So general relativity too may be only an approximation to some deeper, underlying truth. You will return to that idea later in this chapter. For now though, you will take a look at some of the phenomena observed in the Universe today that can only be explained by Einstein's gravitational theory, general relativity.

13.3 Curved space

A consequence of the curved space near to massive objects is that light, or any other electromagnetic radiation, passing close to a massive object will have its path bent. However, saying that gravity bends the path of light is perhaps not the correct way to look at things from the point of view of general relativity. Rather, light continues to travel in a straight line; it is just that space is curved near to a massive object, so (from sufficiently far away) its path appears to deviate.

With the development of **radio astronomy** during the 1960s, this feature of general relativity could be tested accurately for the first time. Radio telescopes were constructed that could determine angular positions to a precision of better than 10^{-6} of a degree. Furthermore, quasars are bright enough at radio wavelengths for them to be detected close to the Sun. A number of measurements of the deflection of radio waves from quasars as the Sun passes in front of them have now been made, and they confirm the predictions of general relativity to a precision of better than one per cent.

A phenomenon related to that described above became apparent in 1979 when astronomers discovered what appeared to be a double quasar. However, the two images had similar spectra and their spectra also had the same redshift as each other. When one of them became brighter, so did the other one. It was soon realised that this was not two quasars, but two images of just one quasar! The situation arises as shown in Figure 13.4, in an effect known as **gravitational lensing**. The idea is that a massive, but dim, galaxy lies almost directly along our line of sight to the distant quasar. Light and other electromagnetic radiation from the quasar travels towards us and is bent by the curvature of space in the vicinity of the intervening massive galaxy. This gives rise to two (or more) images of the lensed object. Since this first discovery, many more such gravitational lensing systems have been found, and some of these are shown in Figure 13.5. A variety of patterns are possible, depending on the mass distribution in the intervening galaxy, and just how closely the distant quasar lines up behind it. As shown in the examples, quadruple patterns and arc-like structures are possible, as well as simple paired images.

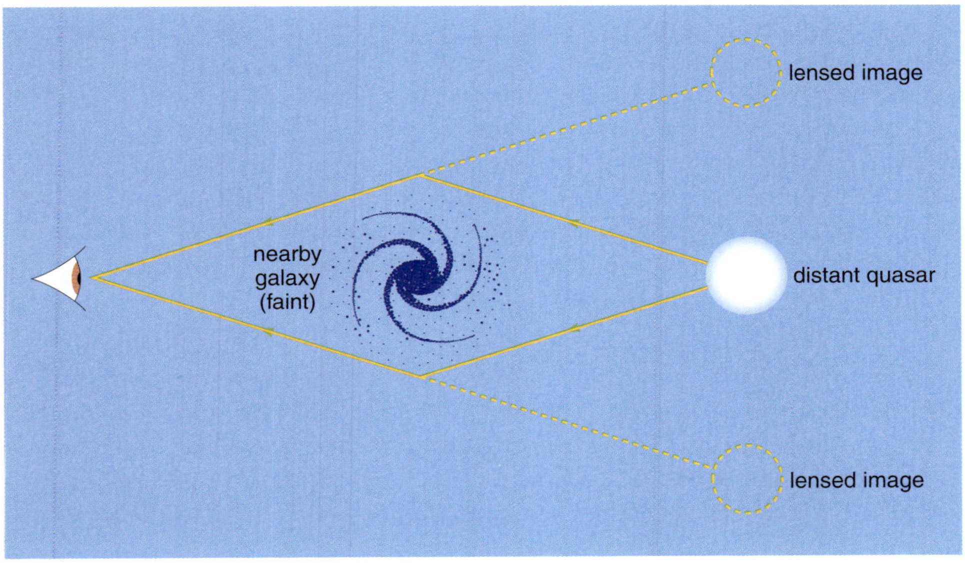

Figure 13.4 The action of a gravitational lens.

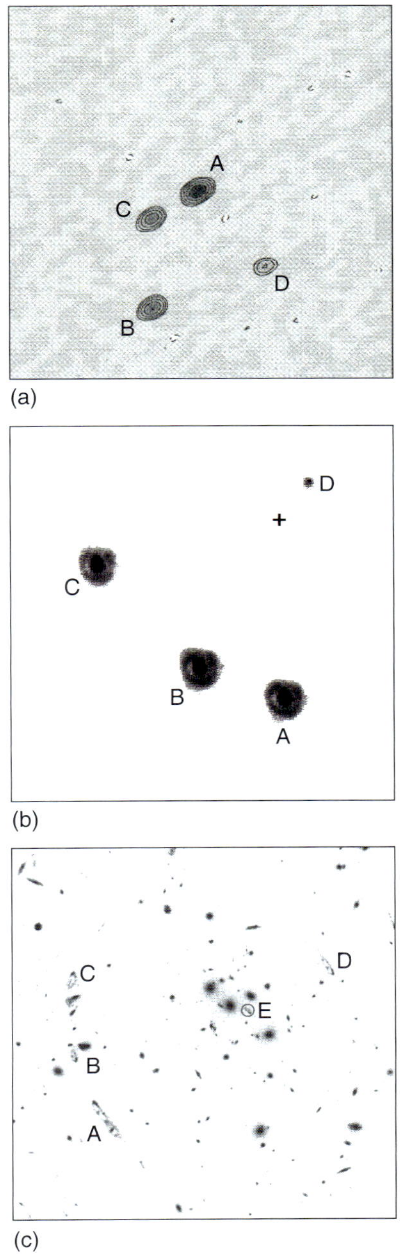

Figure 13.5 Some examples of gravitational lenses. (a) A radio-wave image showing quadruple images of a distant quasar. The intervening faint galaxy lies between the four images labelled A–D. This foreground object shows up on deep-exposure, visible-light images of the same region of sky and lies at a distance of 1.8 Gpc, whilst the multiple-lensed quasar lies much further away. (b) Another quadruple image of a distant quasar, this time in visible light, in an image obtained with the Hubble Space Telescope. The lensing galaxy, whose position is marked with a cross, lies at a distance of about 1.3 Gpc. (c) In this Hubble Space Telescope image, the labels A–E identify 'arcs' that are the result of gravitational lensing by the foreground galaxies – the fuzzy blobs seen in the centre. These arcs are magnified and distorted images of a distant galaxy lying far beyond this cluster of galaxies.

13.4 Gravitational radiation

A final confirmation of Einstein's theory of general relativity is currently on the verge of direct confirmation. It concerns what is known as **gravitational radiation**. When electrically charged particles accelerate (i.e. when they change their speed or direction of motion) they emit electromagnetic radiation. In a similar way, general relativity predicts that when massive objects accelerate (due to either a change in their speed or direction of motion) they will emit gravitational radiation. These gravitational waves can be thought of as ripples in the geometry of space, spreading out from their origin, stretching and squeezing space itself as they pass by. They travel at the speed of light but, in all everyday Earthbound experiences, their effects will be extremely tiny. There are objects in the Universe, however, from which astronomers will soon be able to detect gravitational waves, using detectors currently under construction.

Gravitational radiation may be detected by the effect that it has on objects here on Earth. Current designs for detectors consist of two extremely long beams of laser light (typically a few kilometres in length) orientated at right angles to each other. When a gravitational wave passes through the device it will cause very slight changes in the length of one beam relative to the other. This change in length may be measured by means of the variation produced in an interference pattern formed by combining the two beams. The big problem is that the distortions produced by the passage of the gravitational wave are extremely tiny. The first detectors to be built are capable of measuring relative length changes of one part in 10^{21}. To put this in perspective, if the laser beams in the detectors stretched from the Sun to the nearest star, the equivalent distortion in length would be about the same as the thickness of a human hair!

The next generation of gravitational wave detectors is even more ambitious. A project called LISA (Laser Interferometer Space Antenna) proposes to fly three spacecraft in a triangular formation, each separated from its neighbours by 5 million kilometres, and following the Earth in its orbit around the Sun, about 50 million kilometres behind us (Figure 13.6). The instruments carried by each spacecraft will monitor continuously the separation between them to detect tiny distortions of space caused by passing gravitational waves.

Even before these highly sensitive detectors detect gravitational radiation here on Earth, there is excellent evidence that gravitational radiation does exist, and astronomers can even point to a system where they know it is being produced right now. In 1974, American astronomers Joe Taylor and Russell Hulse discovered a remarkable binary star system: two **neutron stars** orbiting around each other once every eight hours (Figure 13.7). One of the two is also a **pulsar**, producing regular pulses of radio waves as it spins on its axis every 59 ms (milliseconds). This system, which became known as the binary pulsar, has proved to be an ideal test site for general relativity. As the system was monitored over several years, it was discovered that the 8 hour orbital period of the binary pulsar was changing very slightly; the period becomes shorter by 75 μs (microseconds) every year. This implies that the system is losing energy. But where is the energy going? The answer is that the system is steadily giving

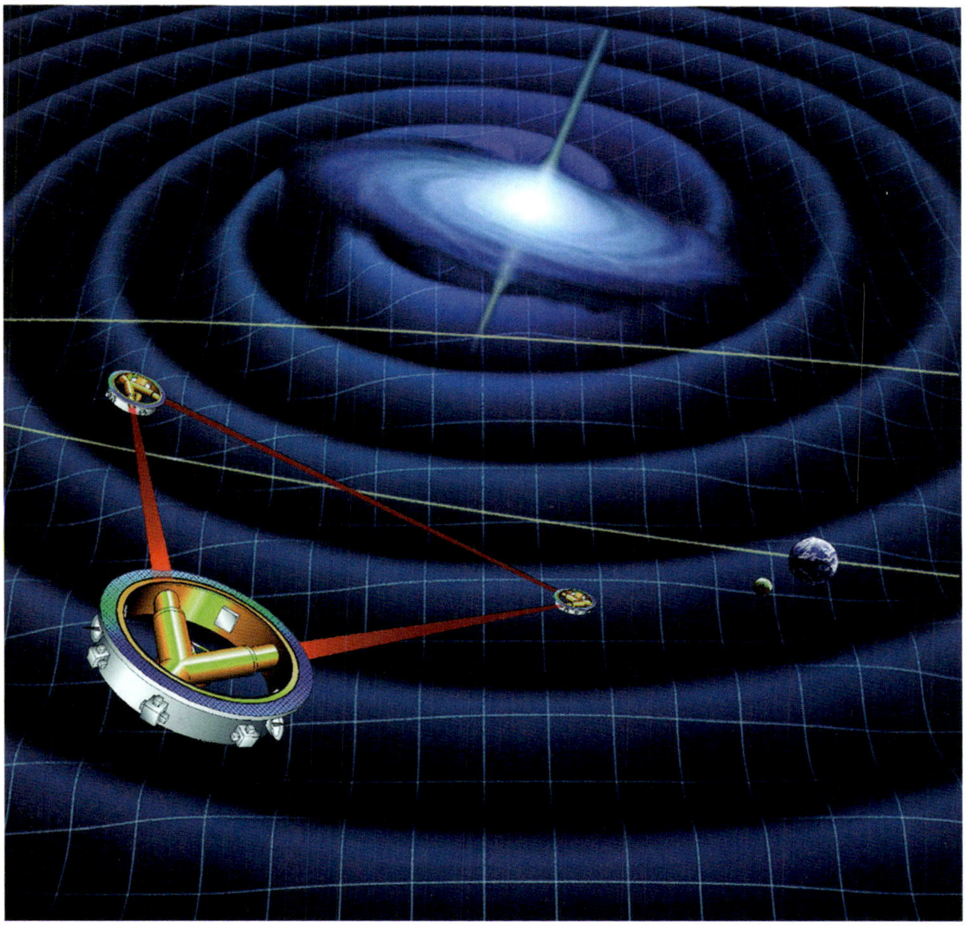

Figure 13.6 A schematic illustration of the planned Laser Interferometer Space Antenna (LISA) for detecting gravitational radiation.

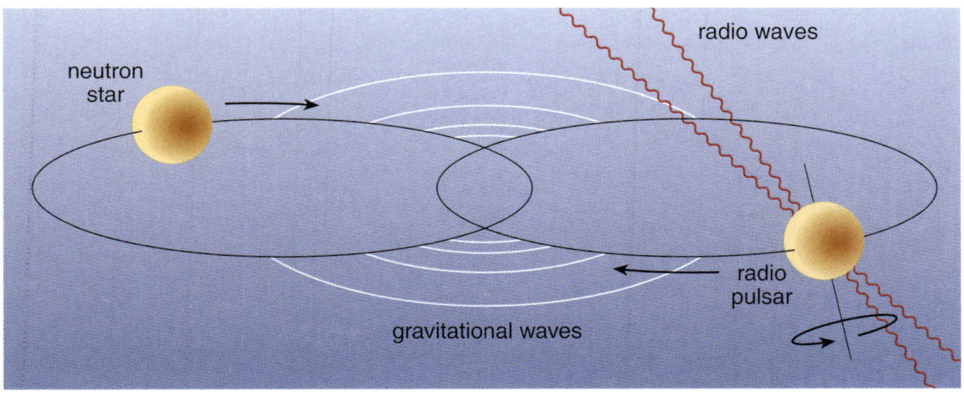

Figure 13.7 Schematic illustration of the binary pulsar system discovered by Hulse and Taylor. They were awarded the Nobel Prize for Physics in recognition of their discovery in 1993.

off gravitational radiation. Calculations showed that the rate of energy loss by gravitational radiation, as predicted by general relativity, ties in exactly with the measured changes in the orbital period of the two neutron stars. Although this gravitational radiation is predicted to be too weak to be detected by the first generation of gravitational wave detectors, there is no doubt that general relativity is being seen in action in this remarkable system. Over the last few years, many more binary pulsars have been discovered and measurements of each of them are in complete agreement with the predictions of general relativity.

13.5 Quantum gravity

The final remarkable feature of gravity is that no one has yet figured out a convincing way of combining quantum physics with general relativity. Nearly a hundred years after Einstein's great work we still do not have a proper relativistic quantum theory of gravity, usually referred to simply as a theory of **quantum gravity**. Einstein made important contributions to quantum physics and he also developed relativity. If anyone deserved to discover a relativistic quantum theory of gravity, it was he. There are two reasons why we still have not got such a theory.

First, there is practically no experimental data on the interplay of gravity and quantum physics, because gravity is so weak at the level of individual particles. The second problem is a conceptual one. It has been mentioned (in Section 4.3) that the quantum physics of atoms involves uncertain velocities and positions for the electrons. In a quantum theory of gravity, somehow the notions of space and time themselves would have to become uncertain. In non-gravitational quantum physics you cannot be sure exactly what you will measure, here and now, or there and then. In a quantum theory of gravity, you would be unsure of what here and now, or there and then, might mean!

Despite these difficulties, it is believed that quanta of gravitational energy exist. They are known as **gravitons** and have zero electric charge and zero mass. It is predicted that gravitons interact with everything; not just material bodies with mass, but also photons and gluons, which have no mass, and even neutrinos, whose masses are as yet poorly known. Moreover, gravitons are predicted to interact with other gravitons too. That makes photons the unique quanta that do not interact significantly with themselves.

While the history of science is full of examples of experiment leading theory, as in the early days of quantum physics, or theory leading experiment, as in the case of the work of Newton, Maxwell and Einstein, it is rare to be stymied on both fronts. In the recent scientific literature there are loads of speculative ideas about what a quantum theory of gravity might involve, one of which will be mentioned, very briefly, in Chapter 14. The difficulty of imagining what a quantum theory of 'fuzzy' space and time might be will set a limit to how far back the history of our Universe can be traced.

Question 13.3

Summing up the differences between Coulomb's law and modern QED theory, you could say: 'Coulomb's law was phrased in terms of a force between electrically charged particles. Modern QED theory describes electromagnetic interactions in terms of the exchange of photons.' Write a similar pair of sentences to describe the differences between Newton's law of gravity and what a quantum theory of gravity might involve.

Activity 13.1 Comparing interactions

We expect this activity will take you approximately 30 minutes.

(a) Having arrived at the conclusion that gravitational interactions do not fit so well into the pattern of description of the three other interactions, you should now update Table 12.1 by completing Table 13.1 below.

Table 13.1 Comparing four interactions.

Interaction	Acts on	Particles experiencing	Quanta
strong			
electromagnetic			
weak			
gravitational			

(b) Finally, to consolidate your understanding of the four fundamental interactions, write a few sentences under each of the following headings in order to compare the interactions with each other: (i) quanta (ii) range (iii) theories (iv) participants (v) strength.

Now look at the comments on this activity at the end of this book.

13.6 Summary of Chapter 13

The gravitational force of attraction is described by Newton's law of gravity. Like Coulomb's law, this is an inverse square law. Newton's law of gravity may be expressed as:

$$F_g = G\frac{m_1 m_2}{r^2}$$

where G is Newton's universal constant of gravitation whose value is $G = 6.7 \times 10^{-11}$ N m^2 kg^{-2}. Near the surface of the Earth, the gravitational force acting on an object of mass m is given by $F_g = mg$ where g is the acceleration due to gravity at the Earth's surface, and has a value of 9.8 m s^{-2}.

Einstein's theory of general relativity describes the interaction between space and the matter within it. When the masses become very large, this theory provides a more accurate description of gravity than does Newton's law. Gravitational lensing of quasars is caused by the curvature of space and is explained by general relativity. General relativity also predicts the existence of gravitational radiation, which is emitted by massive objects that undergo an acceleration. There is good evidence that such radiation is being generated by binary pulsars.

A convincing theory of quantum gravity has yet to be formulated, but it will involve quanta referred to as gravitons which interact with everything.

Chapter 14
Unified theories

> Anything scientists say about the theory begins to sound worryingly like the sort of thoughts that would make you edge away if conveyed to you by a stranger on a park bench.
>
> Bill Bryson, American author (1951–) in *A Short History of Nearly Everything*

The tally of fundamental interactions was completed in Chapters 10–13. According to current reckoning there are *no more than* four fundamental interactions of all matter and radiation: strong, electromagnetic, weak and gravitational. There is great interest, among physicists and cosmologists, in the question: 'Are these four interactions really so distinct, or might they be different facets of some more basic unity of nature?'

The idea of unifying descriptions of force is not new; you saw in Chapter 10 that Maxwell achieved a spectacular unification of electricity, magnetism and light. Recently there has been much activity in investigating a scheme for the unification, at high energies, of weak interactions with electromagnetism. This is known as *electroweak unification* and is the subject of Section 14.1. The consequences will be tested by particle accelerators that are currently under construction, or at the planning stage, and some answers may be known by the time you read this.

Emboldened by this activity, some theorists have suggested a unification of the electroweak theory with QCD, the theory of the strong interaction. This is the subject of Section 14.2, on the so-called *grand unification*. For this theory, testable predictions are harder to come by. Perhaps the Universe itself is our best laboratory.

So as not to leave gravity out of the fold of a unified quantum theory, attempts have been made at what is called *superunification*, which is mentioned briefly in Section 14.3. Such ideas are highly speculative, but may give a clue to the nature of the very early Universe. A schematic picture for this idea of unification is shown in Figure 14.1, which you will revisit over the next few sections.

Please note that Sections 14.1–14.3 are designed to give merely the flavour of enterprises that are the subject of intense current research. Unlike previous parts of the book, they contain matters on which there is no clear consensus. Nevertheless, the questions are likely to endure. Also note that the word 'unification' is used in an *active* sense. This chapter is not about armchair discussions, as to whether things are different or similar. It involves real questions about the behaviour of quarks and leptons, at very high energies, and so it will figure in the account of the history of the Universe given in Chapter 15.

A loose analogy may help. Suppose you are given three substances, at room temperature: A is a solid, B is a liquid and C is a gas. The substances are clearly different, in crucial respects. No amount of words or thought will make them into the same thing. But if you raise the temperature, so that A melts, you may discover that, as liquids, A and B have close similarities, which were

Figure 14.1 A schematic illustration of the unification processes discussed in Chapter 14. At low energies the four fundamental interactions appear to be quite distinct, but at progressively higher energies they become unified.

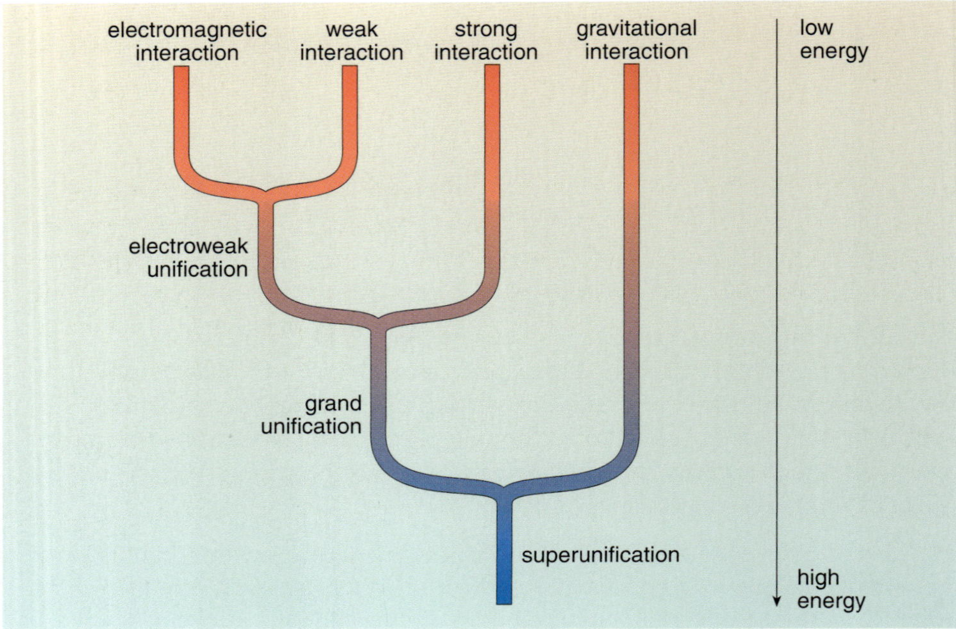

unsuspected at room temperature. That might lead you to a 'unified theory of AB liquids'. Increasing the temperature further, so that both A and B evaporate, you might find strong similarities between A, B and C as gases, and formulate a 'unified theory of ABC gases'. It is a *little* like that with, respectively, the weak, electromagnetic, and strong interactions.

The significance for cosmology is that there was, as you have seen in Chapter 9, an early hot epoch of the Universe when weak and electromagnetic interactions may have been more unified than they now appear. There was probably an even earlier, hotter, epoch when electroweak interactions and strong interactions may have been unified. If so, radically new processes that turn quarks into leptons, and vice versa, may have been in operation, leading to a possible explanation of features of the currently observable, far cooler Universe.

14.1 Electroweak unification

You know, from Chapter 12, that the large masses of the W and Z bosons are responsible for the long lifetime of free neutrons, and the very feeble interactions of low-energy neutrinos from nuclear beta-decay. The similarity in strength of electromagnetic and weak interactions becomes apparent when comparing the interactions of electrons and neutrinos with kinetic energies of around 100 GeV, or greater; well below that energy, there is gross disparity.

A mechanism to explain the high-energy **electroweak unification**, and the lower-energy difference, was proposed in the 1960s by a number of theorists, including Sheldon Glashow, Steven Weinberg and Abdus Salam, who shared the Nobel Prize for Physics in 1979 for their work. The name of one of the other principle scientists involved, Peter Higgs, has become attached to a new particle, the **Higgs boson**, on whose existence the proposal relies. In the mid-1990s construction began on a new particle accelerator, the Large Hadron Collider (LHC) at CERN, near Geneva, shown in Figure 1.3. On this machine, and its recently upgraded neighbour, the Large Electron–Positron collider mark 2 (LEP2), ride the hopes of discovering the Higgs boson, or some new surprise instead.

Just why the Higgs boson should exist is a complicated tale, but the following discussion will give you the general idea. A unified electroweak theory must be able to account for all three quanta involved in the weak interaction (W^+, W^- and Z^0 bosons) as well as the photon that is involved in the electromagnetic interaction. The problem is that the W and Z bosons have mass and interact with each other, whereas photons are massless and do not interact with other photons. Photons cannot have mass, otherwise there would be no such thing as Coulomb's law. Massive quanta – such as the W and Z bosons – cannot produce an inverse square law of force; their effects decrease much faster with distance. The problem then is to develop a theory that explains the existence of four quanta, three of which are different from the other one.

According to the current theory of electroweak unification, there are four so-called 'Higgs fields', one corresponding to each of the W^+, W^- and Z^0 bosons and the photon. Three of these fields 'give mass' to the W and Z bosons. The fourth field does not give mass to the photon, but will be detectable as a true particle – the Higgs boson – with a mass energy of between 100 and 1000 GeV. Therefore, above an energy scale of between 100 and 1000 GeV, the electromagnetic and weak interactions will appear truly unified and merely be different aspects of a single electroweak interaction.

■ Which location on Figure 14.1 corresponds to this energy?

□ It is where the branches representing the electromagnetic and weak interactions join together. Write the value of this energy on Figure 14.1 at the appropriate place.

As you can see, the story of the Higgs boson is a somewhat tangled tale. Nevertheless, it is good science; to get a satisfactory explanation of effects at energies currently available (up to 100 GeV), theorists have been led to predictions at a higher energy (up to 1000 GeV). This outline has been given because it is probable that during the lifetime of this book you will read in the newspapers one or other of the types of headline shown in Figure 14.2.

Figure 14.2 Possible newspaper headlines reporting the hunt for the Higgs boson.

The issue of electroweak unification is important for cosmology; what hope do we have of charting the story of the Universe back to times when the energies were enormously higher than 1000 GeV, if there is a problem in the region between 100 GeV and 1000 GeV?

Question 14.1

(a) Using information from this chapter and from Chapter 12, complete Table 14.1 to summarise information about the mass energies of the quanta involved in electroweak unification.

(b) What is different about photons when compared with the W and Z bosons, and what consequences does this have?

Table 14.1 Quanta involved in electroweak unification.

Quanta	Mass energy/ GeV
photon	
W^+ boson	
W^- boson	
Z^0 boson	
Higgs boson	

14.2 Grand unification

Confirmation of the details of electroweak unification will still leave the strong and gravitational interactions out of the unified fold. Further unification of the forces of nature is an obvious theoretical challenge.

As already mentioned, the strength of electromagnetic interactions (according to QED) *increases*, rather slowly, with the interaction energy involved in the process. The corresponding measure of the strength of the strong interactions (according to QCD) *decreases* with the interaction energy, again rather slowly. This raises an interesting pair of questions: 'At what energy scale might they become equal; and might there be new processes at this energy scale, expressing a **grand unification** of strong and electroweak interactions, leaving only gravity out of the fold?'

The proposed energy scale for grand unification is about 10^{12} times higher than can be achieved with current particle accelerators. As noted in Section 11.2, it is of the order of 10^{15} GeV, as compared with the energies of the order of 10^3 GeV available at the Large Hadron Collider. Such very high energies were probably involved in the early Universe, but they will not be achieved, by human means, on Earth, for the foreseeable future. Above this energy it is predicted that there is a single interaction, characterised by a single strength.

■ Which location on Figure 14.1 corresponds to this energy?

□ It is where the branches representing the electroweak and strong interactions join together. Write the value of this energy on Figure 14.1 at the appropriate place.

The answer to the second question 'What new processes might occur at this energy?' is rather intriguing. The expectation is that there are quite new interactions, involving bosons with mass energies of around 10^{15} GeV. Such hypothetical particles have been named **X bosons**, because nothing is known about them, directly, from experiment. The prediction is that these new interactions allow quarks to change into leptons, matter into antimatter, and vice versa in each case. The reason that such processes have not yet been seen working at energies far below 10^{15} GeV is analogous to the slowness of neutron decay.

Remember that weak interactions, responsible for neutron decay, involve W and Z bosons and are indeed very weak at energies far smaller than the 100 GeV that corresponds to the mass of the W and Z bosons. Processes involving X bosons would similarly be very weak at energies far smaller than the 10^{15} GeV that corresponds to the mass of an X boson, and would be incredibly slow at the energies that are currently observable (up to 1000 GeV). While the behaviour of the hot early Universe would depend crucially on the processes mediated by X bosons, what we observe at lower energies hardly depends on them at all.

However, the new processes *might* show up, very rarely, at lower (i.e. achievable) energies. For example, one effect of such new processes may be that protons are not stable, but decay (eventually). The typical proton lifetime predicted by the Grand Unified Theory is around 10^{33} years. This is immensely longer than the age of the Universe, which is (only!) around 10^{10} years.

■ So how might a proton decay be detected, in a reasonable time, say, of one year?

□ Here the intrinsically random nature of all subatomic processes helps. Starting with 10^{33} protons (equivalent to a mass of more than 1000 tonnes) and waiting for a few years, it might be possible to observe a few decays.

Experiments approaching this sensitivity are currently in progress, and might bear fruit within the lifetime of this book. However, even if evidence in support of grand unification emerges from searches for rare processes, such as proton decay, or (as sometimes happens in science) from less expected quarters, one step will remain in the ambitious attempt to construct a coherent 'theory of everything': the construction of a theory of quantum gravity.

Question 14.2

(a) How many orders of magnitude of energy have you explored in this book, from the study of atomic processes to the quest for a Higgs boson?

(b) How many orders of magnitude are there between the highest energies we can study on Earth and those that might be entailed by grand unification?

14.3 Superunification: strings and branes

As already discussed, in Chapter 13, Einstein transformed the Newtonian theory of gravity as a force into a reinterpretation of space and time. His theory of general relativity involved no quantum physics, though it profoundly modified ideas of space and time. Two of the obstacles to combining it with quantum physics have been mentioned: the difficulty of introducing ideas of uncertainty into the discussion of the properties of space and time; and the lack of laboratory data against which to test such attempts.

A current theory that incorporates quantum gravity involves the descriptions of particles as **strings**, rather than points. Up to now, the various particles (e.g. leptons, quarks and photons) have been described as 'points' that can move through space as time progresses. Each particle can then be characterised by its 'position' in three-dimensional space plus one dimension of time. However, you have seen that particles have other characteristics too, such as electric charge, colour charge, and mass. All of these extra characteristics have to be included to reach a complete description of the particles.

In string theory, particles are replaced by strings that can be open (meaning they have two ends) or closed (meaning they form a loop like an elastic band). Importantly, strings can vibrate, and different modes of vibration (like different notes on a violin string) correspond to particles with different masses or charges (Figure 14.3). Consequently, each type of particle may be represented by a different mode of vibration. One mode of vibration makes the string look like a photon, another makes it look like a muon, yet another mode makes it look like a down quark, and so on. Crucially, there is even a mode of vibration that corresponds to a graviton.

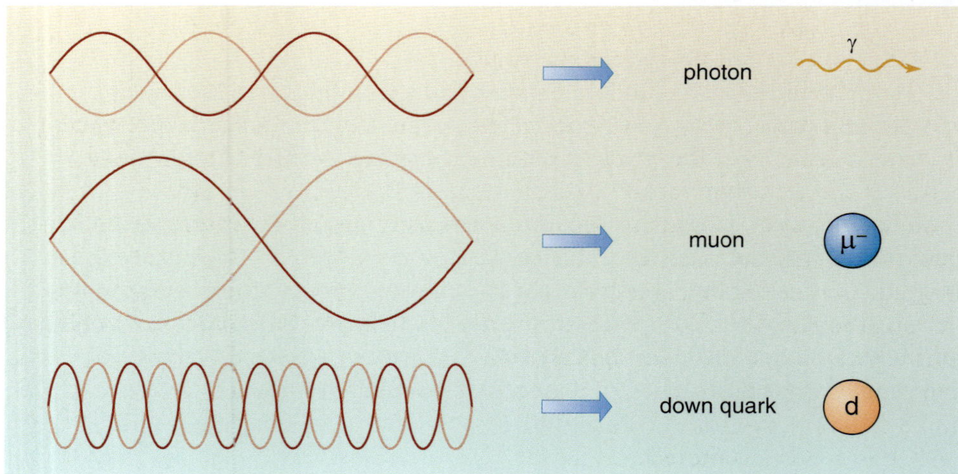

Figure 14.3 Different modes of vibration of strings correspond to different fundamental particles. Of course, these strings are not really like violin strings producing different notes, but the analogy is a useful one.

A remarkable prediction of string theory is that space–time actually has ten dimensions. How absurd! After all, we are only aware of three dimensions of space plus one of time. Where are the other six dimensions predicted by the theory? The explanation is that the other six dimensions are curled up very small so that we do not notice them! Small in this context means around 10^{-35} m, which is about 10^{21} times smaller than an atomic nucleus, so it is no wonder we're not aware of them. The reason why these extra dimensions are needed is that properties such as mass, electric charge and colour charge then correspond to movement, or vibration, of the string within these compact dimensions, just as everyday behaviour such as speed of motion corresponds to movement in the regular three dimensions of space plus one of time.

This is all well and good (even if a little bizarre), but physicists soon discovered not one, but *five* versions of string theory that are consistent and make sense. Some of these involved open strings, some involved closed strings, and some involved both types. If string theory really is the route to a **theory of everything**, why should there be (at least) five different varieties?

A possible answer to this problem came with the discovery that the five string theories are actually different aspects of a single underlying theory, now called **M-theory**. M-theory contains much more than just strings though. It contains other multidimensional objects called **branes**. The word comes from 'membrane', which refers to any two-dimensional structure, like a rubber sheet for instance. In M-theory, however, branes can exist in up to *eleven* dimensions. At present, ideas such as M-theory attract some of the most fertile minds of our age, at an intellectual frontier. Whether they lead to testable predictions remains to be seen. One hope is that the difficulty of achieving **superunification**, of all four interactions, results from there being only one way of doing it, which has not as yet been found.

The typical energy scale at which superunification might occur is known as the **Planck energy**, and it has a value of around 10^{19} GeV (i.e. 10^{28} eV).

■ Which location on Figure 14.1 corresponds to the Planck energy?

□ It is where the branches representing the grand unified interaction and gravitational interaction join together. Write the value of the Planck energy on Figure 14.1 at the appropriate place.

It is at this scale of energy, 10^{16} times higher than the energies that have been studied on Earth, that the standard model for the history of the Universe will begin, in the next chapter. Ultimately, one might hope to derive the entire evolution of the Universe, from the Planck scale of superunification, at 10^{28} eV, through the X-boson scale of grand unification, around 10^{24} eV, down to the Higgs-boson scale of electroweak unification, now under intensive experimental investigation at 10^{12} eV on planet Earth, and from there through the well-charted territory of Chapters 10–12, right down to the 2 eV photons of visible light that carry this message to you. Only time will tell whether or not an understanding of all 28 orders of magnitude lies within the grasp of the human intellect. In either case, it is a glorious project.

Question 14.3

What names are used to describe each of the new types of 'particle' predicted by: (a) electroweak unification; (b) grand unification; (c) current ideas regarding superunification?

Activity 14.1 The home of the World Wide Web

We expect this activity will take you approximately 10 minutes.

The World Wide Web was invented at CERN, the European Organization for Nuclear Research, in Geneva. It is therefore appropriate to visit the CERN website to find out the latest news on the hunt for the Higgs boson. You should go to Activity 14.1 on the course website and follow the link there to find out whether or not the Higgs boson has been found since this book was written.

There are no comments on this activity.

14.4 Summary of Chapter 14

Electroweak unification has been accomplished, on paper, by a theory that gives mass to the W^+, W^- and Z^0 bosons, while leaving the photon massless. A consequence is the prediction of a new, potentially observable, particle: a Higgs boson with a mass energy of between 100 and 1000 GeV. At the time of writing, this particle is being hunted with great tenacity. If it is found, that will be remarkable confirmation of theory; if it is ruled out, that will be a remarkable triumph of experiment. In either case, our understanding of the early Universe will increase.

Extrapolating the observed decrease in strength of the strong force and the observed increase in strength of the electromagnetic force over the well-charted territory up to 100 GeV, it is predicted that grand unification of electroweak and strong interactions occurs at around 10^{15} GeV. Grand unification is expected to involve X bosons that cause proton decay, at a very slow rate, which may be observable on Earth.

At the Planck energy of 10^{19} GeV a theory of quantum gravity is needed, in which the properties of space and time are as indeterminate as those of matter. Such a superunified theory is being sought. One candidate for a superunified theory treats particles as strings, rather than points. Strings exist in ten dimensions, six of which are curled up very small. Different modes of vibration of the strings in these compact dimensions correspond to different masses, electric charge, and colour charge. So strings with different vibration modes become apparent as the various particles we know. Five different string theories have been identified, which is puzzling if they really do represent an ultimate unification. The very latest idea for superunification is known as M- (membrane) theory. This theory incorporates all five of the previously known string theories and predicts that objects called branes exist in eleven dimensions.

PART IV – HOW DOES THE UNIVERSE CHANGE WITH TIME?

> Two things are infinite: the Universe and human stupidity; and I'm not sure about the Universe.
>
> Albert Einstein, German-Swiss-American Physicist (1879–1955)

In this final part of the book, the strands from the previous chapters are woven together to tell the history of our evolving Universe from the instant of the big bang to the present day. The final question posed in Chapter 1 will be answered – how does the Universe change with time?

Chapter 15
The history of the Universe

> In the beginning there was nothing at all. To the north and south of nothingness lay regions of fire and frost.
>
> Snorri Sturleson, Icelandic historian (1179–1241)

> The cosmos is all there is, all there ever was, and all there ever will be.
>
> Carl Sagan, American astronomer (1934–1996)

The two viewpoints expressed above sum up the difficulty in describing the origin of the Universe and perhaps show that we have not really moved very far in the 750 years or so that separate the two statements. This chapter attempts a more detailed description than either of these. The ideas presented are the best explanation we currently have for the reasons why the Universe displays the behaviour that is observed today. In the following sections you will be reminded about all of the fundamental particles, nuclei and atoms that you read about in Part I, as well as the origin of the large-scale behaviour that was presented in Part II. You will also see how the various interactions discussed in Part III have each played a role in determining the present-day structure and contents of the Universe.

15.1 Time, space, temperature and energy

The conventional view of the Universe is that, at the very instant of the big bang, the Universe came into being. There was no 'before' this instant since the big bang marked the creation of time. No location for this event can be specified since the big bang marked the creation of space. All that can be discussed are times after the big bang, and things that happen in the space created as a result of it. This is a difficult concept to visualise; but please stick with it and examine the consequences that follow.

■ What were the conclusions of Chapters 8 and 9 about how the separations between distant objects, and the temperature of the Universe, vary with time? What does this imply about conditions in the early Universe?

□ Chapter 8 concluded that the separations between distant objects are continually *increasing* with time. Chapter 9 concluded that the temperature of the Universe is continually *decreasing* with time. This implies that the early Universe was much *denser* and *hotter* than it is today.

The thread running through this chapter is therefore one of a Universe in which space is expanding, and in which the temperature is falling. In the early part of its history, every time the Universe increased in age by a factor of 100, it also cooled by a factor of ten and distances within the Universe increased by a factor of ten. A consequence of the cooling and expansion is that the mean energy per particle (i.e. the energy available for any reaction to occur) is continually reduced. This has important implications for the ways in which the four fundamental interactions manifest themselves at different epochs.

In Chapters 10–13, you saw that the four fundamental interactions have very different strengths and act on different types of particle. Then, in Chapter 14, you saw the clue to their unification, namely that the strengths of the interactions vary with the energy of their environment. Furthermore, at very high energies, particles can transform into different types – quarks into leptons for instance. So the fact that only quarks feel the strong interaction, whilst leptons do not, is irrelevant at very high energies. Figure 15.1, which is an annotated version of Figure 14.1, shows that at higher and higher energies, first the electromagnetic and weak interactions become unified as energies reach around 1000 GeV. Then the strong interaction becomes unified with the electroweak interaction at an energy of around 10^{15} GeV. Finally, at the very highest energies of at least 10^{19} GeV, gravity too may become unified with all the other interactions.

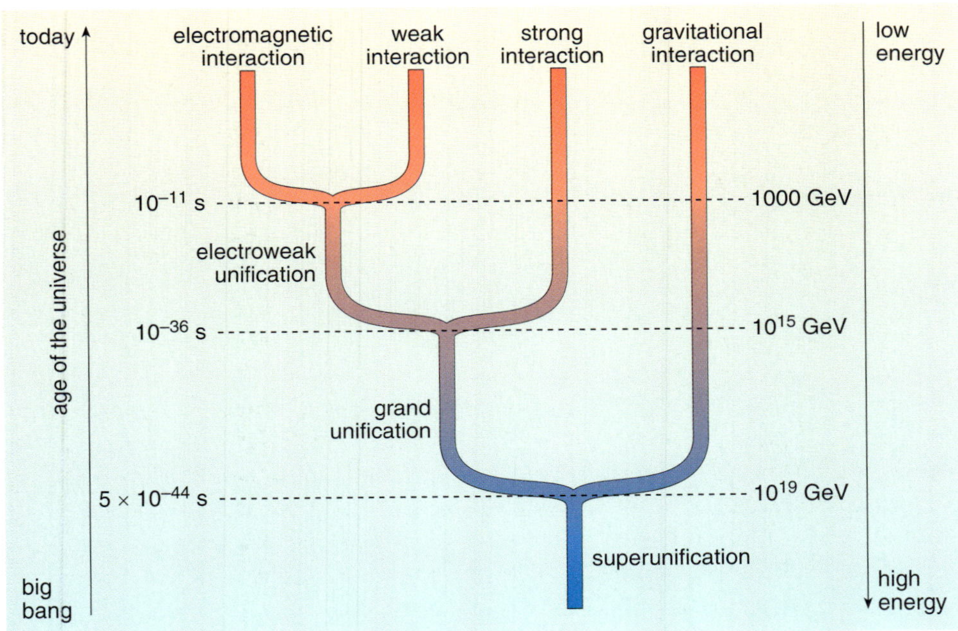

Figure 15.1 The unification of the four fundamental interactions as energy increases towards the bottom of the diagram. As the Universe has aged, so the mean energy of each particle has fallen, and the various interactions have become distinct from one another as the energy has fallen below the thresholds shown. The approximate ages of the Universe when each of these interactions became distinct are also shown.

At the very earliest times, the Universe was extremely hot, the mean energy available per particle was extremely high, and so the unification of interactions discussed in Chapter 14 would have occurred naturally. As the Universe has cooled, the available energy has fallen, and the interactions have in turn become distinct until the current situation is reached in which four different interactions are observed. The relationship between the mean energy of a particle and the temperature of the Universe, and the time at which such energies and temperatures applied, is shown in Figure 15.2.

You will appreciate from Figure 15.2 that the rest of Chapter 15 will necessarily refer to incredibly short intervals of time after the big bang (notice how far along the graph 1 second appears). Many important processes took place when the Universe was significantly less than 1 s old, when the energy available for processes in the Universe was extremely high. In fact, most of the important processes were completed by the time the Universe was only a few minutes old!

For each of the intervals under discussion in Sections 15.2–15.7 (and shown in Figure 15.2) the time, temperature and energy ranges are given. As you read through these sections, you can imagine yourself travelling down the line of the graph in Figure 15.2.

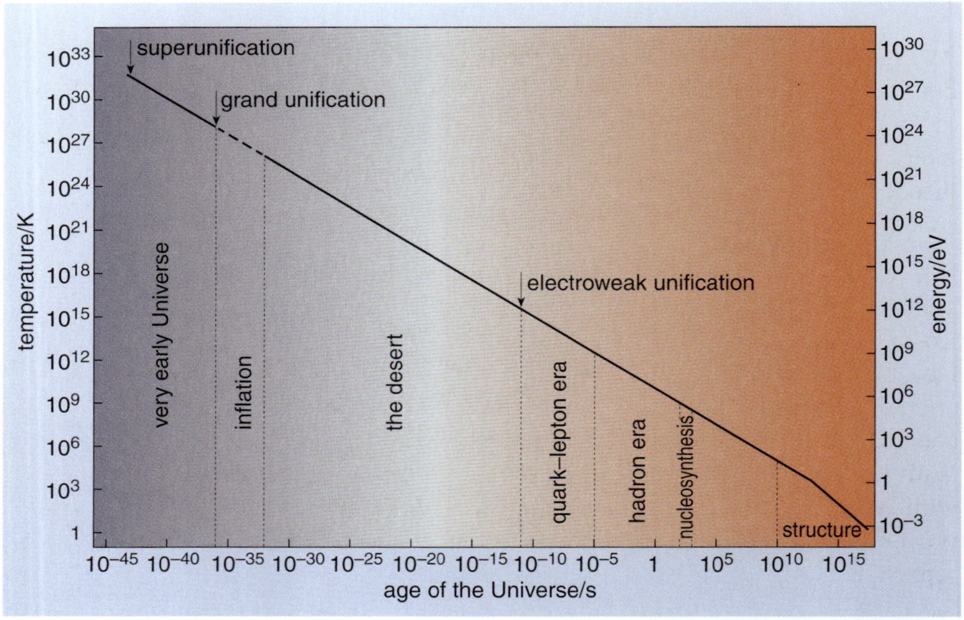

Figure 15.2 The temperature and mean energy per particle of the Universe at different times in its history. The divisions of this graph are described in Sections 15.2–15.7. Notice that the axes are shown on logarithmic scales to accommodate the vast ranges of energy, temperature and time that are necessary. You will find it useful to refer back to Figures 15.1 and 15.2 as each successive stage in the history of the Universe is introduced.

15.2 The very early Universe

Time: $<10^{-36}$ s

Temperature: $>10^{28}$ K

Energy: $>3 \times 10^{15}$ GeV (i.e. $>3 \times 10^{24}$ eV)

At the very earliest times in the history of the Universe, it can only be presumed that a superunification of the four interactions existed. As you have seen, no reliable theory of superunification is yet available, so *nothing* can be said about the contents or behaviour of the Universe in its earliest moments. It may even be that the concept of 'time' itself had no meaning until the Universe had cooled below a certain threshold.

The first stop on the tour where anything can be said is at about 3×10^{-44} s after the big bang – an epoch known as the **Planck time**. By this time, the mean energy per particle in the Universe had fallen to around 10^{19} GeV (the Planck energy that you met in Chapter 14). This is the energy at which the gravitational force on an individual particle has roughly the same strength as its other interactions. An idea of the typical size scale of the Universe can be gained by thinking about how far a photon of light could have travelled during this period. By the time the Universe was 3×10^{-44} s old, a beam of light travelling at 3×10^{8} m s^{-1} could have travelled a distance of only about 10^{-35} m. This tiny dimension is referred to as the **Planck length**. As noted in Section 14.3, this is the sort of scale on which the hidden dimensions involved in M-theory are supposed to be curled up very small.

■ How does the Planck length compare with the typical size of an atomic nucleus?

□ The typical size for an atomic nucleus is about 10^{-14} m. So the Planck length is around 10^{21} times smaller than an atomic nucleus. (It is as many times smaller than a nucleus as a nucleus is smaller than the Earth!)

As Figure 15.1 shows, at or around the Planck time, it is supposed that gravitational interactions became distinct from a grand unified interaction that included the three effects seen today as the electromagnetic, strong and weak interactions. In order to describe the gravitational interactions at these times, a theory of quantum gravity is required. However, as you saw in Section 13.5, no such theory is yet available.

The temperature, and therefore the mean energy per particle, was far higher at this time than can be recreated in particle accelerators here on Earth. Cosmologists and particle physicists can therefore only speculate on what might have occurred in the very early Universe. The best guess is that pairs of matter and antimatter particles of all types were spontaneously created out of pure energy, which can be thought of as a 'sea' of photons filling the entire Universe. With equal spontaneity, pairs of matter and antimatter particles also combined with each other again to produce photons. As you saw in Section 2.2, the overall process of pair creation (left to right) and annihilation (right to left) can be represented as:

$$\text{photons} \rightleftharpoons \text{particle} + \text{antiparticle} \qquad (15.1)$$

At the temperatures existing in the Universe today, reactions such as this proceed preferentially from right to left. However, at the temperatures applying in the early Universe, the reactions proceeded in both directions at the same rate, for all types of particle. A stable situation was reached in which the rates of pair creation and annihilation exactly balanced, and equal amounts of matter/antimatter and radiation were maintained.

As well as the familiar quarks and leptons, if the Grand Unified Theory discussed in Chapter 14 is correct, then this is when the particles known as X bosons would also have been in evidence. These particles are the quanta of the grand unified interaction and are suggested as a means of *converting* between quarks and leptons, or between matter and antimatter.

The next stop in time is at about 10^{-36} s after the big bang when the Universe had a temperature of about 10^{28} K. This temperature marks the energy at which the strong interactions became distinct from the electroweak interactions (Figure 15.1).

■ How long after the Planck time did the strong and electroweak interactions become distinct?

□ Be careful here!

$10^{-36}\,\text{s} - (3 \times 10^{-44}\,\text{s}) = (1 \times 10^{-36}\,\text{s}) - (0.000\,000\,03 \times 10^{-36}\,\text{s}) = 0.999\,999\,97 \times 10^{-36}\,\text{s}.$

So, to all intents and purposes, the strong and electroweak interactions became distinct about 10^{-36} s after the Planck time.

It should be emphasised that there is some disagreement and uncertainty about the exact processes that occurred at this extremely early period in the history of the Universe, but the story outlined above is the best guess at what may have actually occurred. Before proceeding with the trip through time, you will pause for a moment to examine a quite remarkable event that seems to have happened just after the strong and electroweak interactions became distinct. The event has profound consequences for the nature of the Universe today.

15.3 Inflation

Time: 10^{-36} s to 10^{-32} s

Temperature: rapidly changing

Energy: rapidly changing

When talking about the Universe, there is an important distinction that the present discussion has, up until now, largely ignored. First, there is the entire Universe and this may be infinite in size as mentioned in Chapter 8. By implication, therefore, it makes no sense to put a value on the 'size' of the entire Universe. But there is also what may be called the *observable* Universe, which is that part of the Universe that it is theoretically *possible* to be observed from Earth. A value for the size of this finite region *can* be calculated.

■ Why should there be a limit to how far we can see?

□ The speed of light is a cosmic speed limit – nothing can travel any faster. So, the only part of the Universe that is now observable is that fraction of it from which light has had time to reach us since the Universe began, about 14 billion years ago (Section 8.6).

One might naturally expect that the radius of the currently observable Universe is therefore equal to the maximum distance that light can have travelled since the Universe began, as illustrated in Figure 15.3. Notice that, although light can travel from the edge of the sphere shown in Figure 15.3 to the centre within the age of the Universe, light *cannot* travel from one edge to the opposite edge (across the diameter of the sphere) within the age of the Universe – the Universe is simply not old enough!

When trying to understand the large-scale structure of the Universe that is observed today, one of the most intriguing problems is that the Universe is so uniform. The results from the COBE satellite (Section 9.2) showed that one part of the Universe has exactly the same temperature, to a precision of better than one part in ten thousand, as any other part of the Universe. The expansion rate of the Universe in one direction is observed to be exactly the same as that in any other direction too. In other words, the observable Universe today is seen to be incredibly *uniform*. At 10^{-36} s after the big bang, when things were far closer together than they are now, the physical conditions across the Universe must therefore have been identical to an extremely high level of precision. Yet, according to conventional physics, there has not been time for these regions of space to ever 'communicate' with one another – no light signals or any other form of energy could travel from one to the other and smooth out any irregularities.

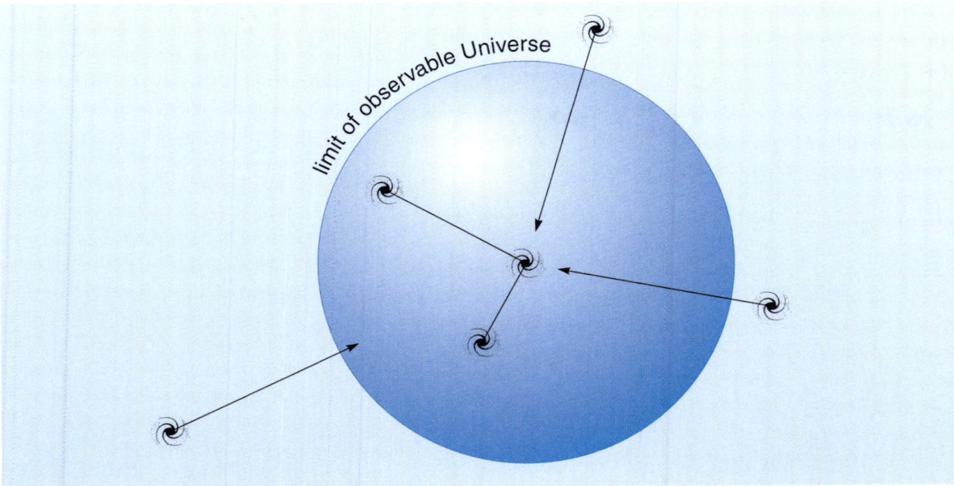

Figure 15.3 The size of the observable Universe. Imagine that the Earth lies at the centre of the circle (really a sphere in three dimensions), with a radius equal to the distance that light can travel since the Universe began. Then light from galaxies lying within the circle has had time to reach us, but light from galaxies lying outside the circle has not had sufficient time to reach us since the Universe began. Such galaxies are simply not observable at the current time. Furthermore, light from galaxies at one side of the circle has not had time to reach galaxies on the other side of the circle.

In 1981, the American physicist Alan Guth suggested that, in the early history of the Universe at times between about 10^{-36} s and 10^{-32} s after the big bang, the Universe underwent a period of extremely rapid expansion, known as **inflation**. During this time, distances in the Universe expanded by an extraordinary factor – something like 10^{50} has been suggested although this could be a vast underestimate!

It is believed that inflation may be caused by the way in which the strong and electroweak interactions became distinct. The exact mechanism by which inflation occurred is not important here, but there are many consequences of this theory. The most important for the present discussion is that the region that was destined to expand to become the currently observable Universe originated in an extremely tiny region of the pre-inflated Universe. This tiny region was far smaller than the distance a light signal could have travelled by that time and so any smoothing processes could have operated throughout the space that now constitutes the observable Universe. The problem of the uniformity of the microwave background and the uniform measured expansion then goes away.

Non-uniformities may still be out there, but they are far beyond the limits of the observable Universe – and always will be, according to current theories. Because we cannot ever hope to see beyond this barrier, we can have no knowledge whatsoever of events that occurred *before* inflation, since any information about such events is washed out by the rapid increase in scale. Inflation serves to hide from us any event, process or structure that was present in the Universe at the very earliest times.

If you are thinking that the inflation theory contains some pretty bizarre ideas – you are right, it does – but it is the most promising theory that currently exists for one of the earliest phases in the history of the Universe. No more will be said about it here, but now the story will be picked up again after the Universe has completed its cosmic hiccup. The strong and electroweak interactions have now become distinct and the X bosons have therefore disappeared.

As the matter and antimatter X bosons decayed, they produced more quarks, antiquarks, leptons and antileptons – so adding to the raw materials from which the material contents of the Universe were later built. Using X to represent a matter X boson and $\overline{X}$ to represent an antimatter X boson, the types of reaction that are believed to have occurred are:

$$X \rightleftharpoons \text{quark} + \text{quark} \tag{15.2a}$$

$$X \rightleftharpoons \text{antiquark} + \text{antilepton} \tag{15.2b}$$

$$\overline{X} \rightleftharpoons \text{quark} + \text{lepton} \tag{15.2c}$$

$$\overline{X} \rightleftharpoons \text{antiquark} + \text{antiquark} \tag{15.2d}$$

All six flavours of quark (u, d, c, s, t, b) and all six flavours of lepton (e^-, μ^-, τ^-, ν_e, ν_μ, ν_τ) that you met in Chapter 7 were produced at this time, along with their antiparticles. Notice, however, that matter and antimatter X bosons can each decay into *either* matter *or* antimatter particles. This will be important later on in the story.

15.4 The quark–lepton era

Time: 10^{-11} s to 10^{-5} s

Temperature: 3×10^{15} K to 3×10^{12} K

Energy: 1000 GeV to 1 GeV

During the time interval 10^{-32} s to 10^{-11} s, i.e. for the 10^{-11} seconds or so after inflation, nothing new happened in the Universe. It merely carried on expanding and cooling, but no new processes took place. The desert (as it is known) came to an end when the Universe reached a temperature of about 3×10^{15} K, and this is where the next stage in our history begins. At this point, the mean energy per particle was around 1000 GeV and the electromagnetic and weak interactions became distinct. As you saw in Chapter 14, the energies corresponding to this transition are becoming attainable in experiments here on Earth. So it could be argued that all particle reactions that models propose after the first 10^{-11} s of the history of the Universe are now *directly* testable in Earth-based laboratories.

By 10^{-11} s after the big bang, the X bosons had long since decayed in reactions like those shown in Equation 15.2, but the temperature of the Universe was still too high for the familiar baryons (protons and neutrons) to be stable. The Universe contained all types of leptons, quarks, antileptons and antiquarks as well as photons. In fact, there would have been approximately equal numbers of particles and antiparticles at this time – but note that word *approximately* – you will look at the implications of this in a moment. There would also have been equal amounts of radiation (photons) and matter/antimatter (particles or antiparticles).

The next stage of the story is to consider how and when the original mixture of all types of quarks and leptons, that were present when the Universe was 10^{-11} s old, gave rise to the Universe today, which seems to be dominated by protons, neutrons and electrons.

■ In particle accelerators, how much energy is required in order to create a particle and antiparticle of a given mass?

□ Broadly speaking, an amount of energy equivalent to (or greater than) the combined mass energy of the particle and antiparticle concerned needs to be supplied. For example, the mass energy of an electron is about 510 keV, so to create an electron–positron pair, at least 2×510 keV ~ 1 MeV of energy must be available.

In the early Universe, when the mean energy per particle was greater than the mass energy of a given particle plus antiparticle, those particles and antiparticles existed in abundance, and survived in equilibrium with radiation. When the mean energy per particle dropped below this value, annihilations became more likely than pair creations, and so the number of particles and antiparticles of a given type declined.

Massive quarks and leptons also decay into less massive ones, and these decays became more likely as the available energy fell. Broadly speaking, when the temperature of the Universe fell below that at which the mean energy per particle was similar to the mass energy of the particles concerned, then the particles decayed into other less massive particles.

So, by the time the Universe had cooled to a temperature of 3×10^{12} K, equivalent to a mean energy per particle of about 1 GeV, when the Universe was 10^{-5} s old, several important changes had taken place. First, many of the tauons and antitauons, muons and antimuons had decayed into their less massive lepton counterparts, namely electrons and positrons. The only leptons that remained in the Universe in any significant number were therefore electrons and neutrinos (with their antiparticles in approximately equal numbers). Similarly, the massive second and third generation quarks (strange, charm, top and bottom) had mostly decayed into their less massive first generation counterparts (up and down), via a range of weak interactions involving W bosons (Section 12.2). In each case, quarks changed flavour with the emission of a lepton–antilepton pair.

All types of quarks and antiquarks also underwent mutual annihilations – with a particularly crucial result. In discussing the relative numbers of particles and antiparticles earlier, the phrase *approximately* equal was used deliberately. If the Universe had contained *exactly* equal numbers of quarks and antiquarks, then these would have all annihilated each other, leaving a Universe that contained no baryons – so no protons and neutrons – no atoms and molecules – no galaxies, stars, planets or people. Clearly that is *not* what we observe around us!

In fact, the Universe now seems to consist almost entirely of matter (rather than antimatter) in the form of protons, neutrons, electrons and electron neutrinos, plus photons. And there are believed to be roughly ten billion photons for every baryon (proton or neutron) in the Universe today. This implies that, just before the quark–antiquark annihilations took place, for every ten billion antimatter quarks there must have been *just over* ten billion matter quarks. Running the Universe forward from this point, for every ten billion quarks and ten billion antiquarks that annihilated each other producing photons, a few quarks were left over to build baryons in order to make galaxies, stars, planets and people.

Why did the Universe produce this slight imbalance of matter over antimatter? Maybe it was just 'built-in' from the start, like any other constant of nature? This is rather unsatisfactory to many cosmologists and particle physicists who prefer to believe that the imbalance arose *after* the Universe had got started. It has been suggested that the decays of X bosons into quarks and leptons *may* slightly favour the production of matter particles over antimatter particles. As you saw in Equations 15.2a–d, a matter or antimatter X boson can decay into *either* matter particles or antimatter particles. So, if there is an imbalance in the rates, starting with equal numbers of matter and antimatter X bosons *will not* lead to the production of equal numbers of matter and antimatter quarks and leptons. Such matter–antimatter asymmetry has actually already been observed with experiments on Earth that measure the decay of particles called K mesons. Of the two possible routes for this reaction, one is favoured over the other by seven parts in a thousand. Perhaps something similar, of the order a few parts in ten billion, occurs with X boson decays? The answer to this question is not yet known – but it is a rather important one, since without it none of us would be here to discuss the matter! It is a rather humbling thought that the existence of the entire current contents of the Universe may be the result of an imbalance in the rates of two decay reactions by a few parts in ten billion.

15.5 The hadron era

Time: 10^{-5} s to 100 s

Temperature: 3×10^{12} K to 10^9 K

Energy: 1 GeV to 300 keV

From the time that the temperature fell to about 3×10^{12} K, at about 10^{-5} s after the big bang, stable baryons (protons and neutrons) began to form from the up and down quarks that remained after the annihilation of matter and antimatter.

■ How does the mean energy per particle at 10^{-5} s compare with the mass energy of a proton or neutron?

□ Protons and neutrons have a mass energy of about 1 GeV, which is similar to the mean energy per particle in the Universe at this time.

This is why confinement of quarks became important from this time onwards. Before 10^{-5} s after the big bang, there had been sufficient energy available for up and down quarks to escape to distances significantly larger than the dimensions of a proton or neutron. After this time, no such escape was possible. Since a proton is composed of two up quarks and a down quark and a neutron is composed of two down quarks and an up quark, equal numbers of up and down quarks therefore led to an equal number of protons and neutrons emerging from this process. To recap on the contents of the Universe at this time, there were about ten billion photons, electrons, positrons, neutrinos and antineutrinos for every single proton or neutron in the Universe.

■ Why had the electrons and positrons not yet mutually annihilated each other?

□ The mass energy of an electron or positron is about 510 keV, and the mean energy per particle was still much higher than the ~1 MeV required to create a pair of them. So electrons and positrons were still in equilibrium with photons, undergoing both annihilation and pair creation reactions at the same rate.

As soon as baryons had formed, weak interactions took over, with protons and neutrons existing in equilibrium governed by the following processes:

$$e^+ + n \rightleftharpoons p + \bar{\nu}_e \qquad (15.3a)$$

$$\nu_e + n \rightleftharpoons p + e^- \qquad (15.3b)$$

So neutrons converted into protons by reacting with either positrons or electron neutrinos, whilst protons converted into neutrons by reacting with either electron antineutrinos or electrons.

■ At the quark and lepton level, how may the two reactions in Equation 15.3 be represented?

□ Bearing in mind the quark composition of a proton and a neutron, each of the reactions involve conversions between a down quark and an up quark as shown in Equations 15.4a and b:

$$e^+ + d \rightleftharpoons u + \bar{\nu}_e \qquad (15.4a)$$

$$\nu_e + d \rightleftharpoons u + e^- \qquad (15.4b)$$

Each of these reactions is a weak interaction because it involves quarks and leptons changing flavour. In each case, a W boson is involved.

With plenty of energy available, the transitions from neutron to proton and from proton to neutron proceeded at the same rate. Since there were as many neutrinos as electrons, and as many antineutrinos as positrons, the numbers of neutrons and protons in the Universe remained equal, at least initially. However, this situation did not continue. As noted in Section 12.3, the mass of a neutron is slightly higher than that of a proton. As a consequence of this, the reactions in which a proton converted into a neutron became slightly less likely to happen as the energy fell, because they required more energy than those in which a neutron was converted into a proton. As the Universe cooled, this difference in the rates of the two processes became more pronounced, and protons began to outnumber neutrons for the first time.

As the Universe cooled still further, another reaction became important for the neutrons and the protons: as you saw in Section 12.3, isolated neutrons decay into protons. This additional process, again governed by the weak interaction, added to the dominance of protons over neutrons in the Universe:

$$n \longrightarrow p + e^- + \bar{\nu}_e \qquad (15.5)$$

Once the Universe was 0.1 s old, the weak interactions described by the reactions in Equations 15.3 and 15.4 became too slow, and neutrinos virtually ceased to have any further interaction with the rest of the Universe – ever! The ratio of protons to neutrons continued to rise as a result of neutron decay, and was only halted (Section 15.6) when the neutrons became bound up in atomic nuclei where they became essentially immune from decay. As you saw in Section 12.3, if the typical lifetime of the neutron (about 10 minutes) were much shorter than it in fact is, then all neutrons would have decayed into protons long before they could become confined inside nuclei.

It is worth pausing for a moment to try and comprehend just how short a time in the history of the Universe has elapsed at this point. Snap your fingers and that is about the same duration as the entire history of the Universe up to the instant now being discussed (i.e. about one-tenth of a second). Inflation, the breakdown of superunification, grand unification and electroweak unification, the formation of quarks and leptons, the subsequent decay of the massive particles, and the binding of up and down quarks into protons and neutrons – all occurred in the first tenth of a second after the big bang – literally in the blink of an eye ...

When the Universe was about 10 s old, and the mean energy per particle was about 1 MeV, a final important event for the matter contents of the Universe occurred. The remaining primordial electrons and positrons mutually annihilated, producing yet more photons, but leaving the excess one-in-ten-billion electrons to balance the charges of the primordial one-in-ten-billion protons and ensure that the Universe has a net electric charge of zero.

15.6 Primordial nucleosynthesis

Time: 100 s to 1000 s

Temperature: 10^9 K to 3×10^8 K

Energy: 300 keV to 100 keV

As the temperature continued to decrease, protons and neutrons were able to combine to make light nuclei. This marked the beginning of the period referred to as the era of primordial **nucleosynthesis** (which literally means 'making nuclei').

You have already encountered a set of nucleosynthesis reactions in Section 6.6, namely the PP chain that occurs in the core of the Sun and other stars. However, the first step in the PP chain, in which two protons react together to form a deuterium nucleus, is a very slow process and did *not* occur to any great extent in the early Universe. On average, an individual proton has to wait more than ten billion years before such a reaction happens and there was simply not enough time in the early Universe for this to contribute significantly to nucleosynthesis.

Instead, the first such reaction to become energetically favoured was that of a single proton and neutron combining to produce a deuterium nucleus, with the excess energy carried away by a gamma-ray photon:

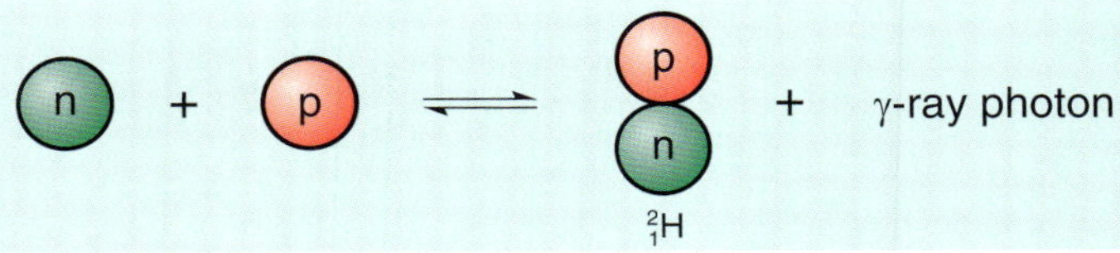

(15.6)

This reaction to produce deuterium does not occur as part of the PP chain in the Sun, as the Sun contains no free neutrons.

■ What is deuterium?

□ Recall from Section 2.1 that deuterium is an *isotope* of hydrogen. Whereas normal hydrogen nuclei consist simply of a proton, deuterium nuclei (sometimes called 'heavy hydrogen') contain a proton and a neutron.

At high temperatures (greater than 10^9 K), there were a lot of high-energy photons in the Universe so this reaction was favoured to go from right to left. As a result, deuterium nuclei were rapidly broken down. However, as the temperature fell below 10^9 K when the Universe was about 100 s old, deuterium production was favoured. Virtually all of the remaining free neutrons in the Universe were rapidly bound up in deuterium nuclei, and from then on other light nuclei formed. One of the reactions that occurred was:

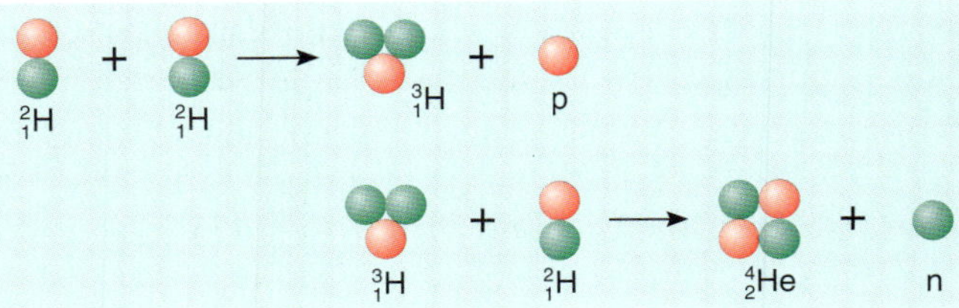

(15.7)

■ What is the nucleus represented by ^3_1H above?

□ This represents a nucleus of another isotope of hydrogen (called tritium) which contains two neutrons and one proton.

This shows that two deuterium nuclei react together to form a nucleus of tritium with the ejection of a proton. The tritium nucleus immediately reacts with another deuterium nucleus to form a nucleus of helium-4 with the emission of a neutron. The proton and neutron produced in the two reactions above can combine to form another deuterium nucleus, so the *net* result of this set of reactions is that two deuterium nuclei are converted into a single nucleus of helium-4.

Other more massive nuclei were also made as follows (the first two of these reactions are similar to the later stages of the PP chain, which you saw in Section 6.6):

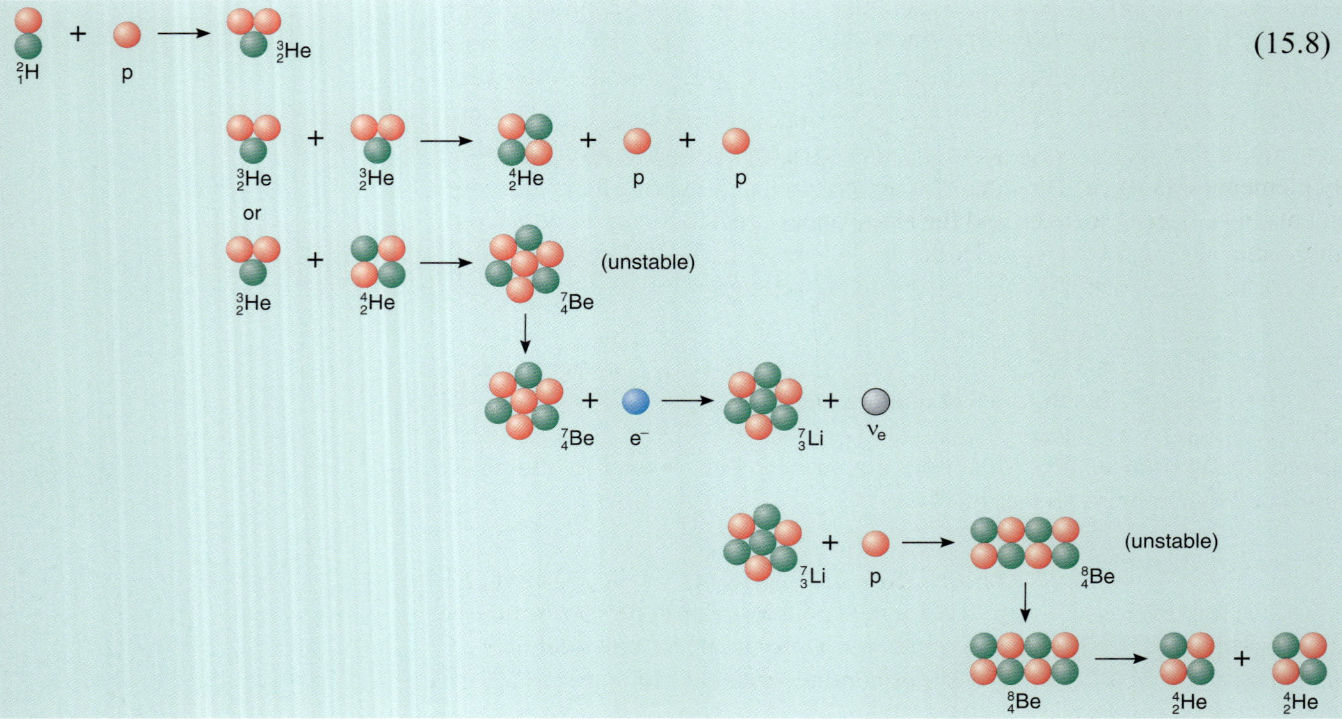

(15.8)

This shows that deuterium nuclei react with protons to make nuclei of helium-3. These can then either react with other helium-3 nuclei to make helium-4 plus more protons or with nuclei of helium-4 to make beryllium-7. Nuclei of beryllium-7 are unstable and immediately capture an electron to form lithium-7 with the emission of an electron neutrino. Lithium-7 nuclei can react further with a proton to create nuclei of beryllium-8, but these too are unstable and immediately split apart into a pair of helium-4 nuclei. The end products of the four reactions are nuclei of helium-3, helium-4 and lithium-7, with the vast majority ending up as helium-4.

Nuclei with a mass number greater than seven did not survive in the early Universe. This is because there are no stable nuclei with a mass number of eight. The binding energy of a beryllium-8 nucleus is –56.5 MeV, which is slightly *more* than the binding energy of two helium-4 nuclei (–28.3 MeV each). Consequently, it is energetically favourable for beryllium-8 nuclei to split apart into two helium-4 nuclei, releasing 0.1 MeV of energy. The reactions that by-pass this bottleneck take much longer than the few minutes that were available for nucleosynthesis at this time. (Remember, the timespan here was around 15 minutes, between when the Universe had an age of 100 and 1000 s.) Before more advanced reactions could occur, the Universe cooled too much to provide the energy necessary to initiate them.

The ratio of protons to neutrons had, by this time, reached about seven protons for every one neutron. Because the neutrons were bound up in nuclei, they no longer decayed, and the ratio remained essentially fixed from here on. The vast majority of the neutrons ended up in nuclei of helium-4. Only very tiny fractions were left in deuterium, helium-3 and lithium-7 nuclei, since the reactions to produce them were far more likely to continue and produce helium-4 than they were to halt at these intermediate products.

By the time the Universe had cooled to a temperature of about 3×10^8 K after 1000 s, the particles had insufficient energy to undergo any more reactions. The era of primordial nucleosynthesis was at an end, and the proportion of the various light elements was fixed. The rates of reaction to form helium and the other light elements have been calculated and the abundances predicted may be compared with the abundances of these nuclei that are observed in the Universe today. There is close agreement between theory and observation.

The close agreement between the theoretically predicted abundances of the light elements and the observed abundances in the Universe today is the third major piece of evidence, alongside the cosmic microwave background and the Hubble expansion, in favour of the hot big bang model for the origin of the Universe (Section 9.3).

At an age of 1000 s, the Universe reached a state where its matter constituents were essentially as they are today. There are about ten billion photons for every baryon (proton and neutron), and about seven protons and electrons for every one neutron. Neutrinos and antineutrinos continue to travel through the Universe unhindered by virtually anything they encounter.

Worked Example 15.1

Assume that the Universe contains one neutron for every seven protons, and that all the neutrons are today bound up in nuclei of helium-4.

(a) What are the relative numbers of hydrogen and helium nuclei in the Universe?

(b) What are the relative percentages, by mass, of hydrogen and helium in the Universe?

(a) One way to calculate the answer is as follows. Imagine that you have a box containing 14 protons and two neutrons – the 7 : 1 ratio mentioned in the question. If a nucleus of helium-4 is made from two protons and two neutrons, there will be 12 protons remaining in the box, each of which can be considered as a hydrogen nucleus. Therefore there are 12 hydrogen nuclei for every one helium-4 nucleus in the Universe.

(b) Taking the mass of a helium-4 nucleus to be four units, and that of a hydrogen nucleus to be one unit, the relative masses of the helium-4 and hydrogen in the box are 4 and 12, respectively. The fraction of the mass in the box due to helium-4 is therefore:

$$\frac{4}{4+12} = 0.25 \text{ or } 25\%$$

and that due to hydrogen is:

$$\frac{12}{4+12} = 0.75 \text{ or } 75\%$$

In fact, the actual proportion by mass of helium-4 that is predicted to have come out of the big bang according to the most recent calculations is 24.8%. The other 75.2% of the mass of nuclei created in the early Universe was virtually all hydrogen. The proportions by mass of deuterium and helium-3 were only about 1 in 40 000 and 1 in 100 000 respectively and the proportion by mass of lithium-7 was about 1 in 2 billion.

15.7 Structure in the Universe

Time: 10^{10} s to 4×10^{17} s (300 years to 14 billion years)

Temperature: 10^5 K to 2.73 K

Energy: 30 eV to 7×10^{-4} eV

As the Universe cooled still further, nothing much happened for a few hundred years (between 1000 s and 10^{10} s). As the mean energy per particle fell below a few tens of eV, so electrons began to combine with nuclei to form neutral atoms.

Gradually, as this electrically neutral matter accumulated, gravity began to take over as the dominant force operating in the Universe. Slight variations in the amount of matter and radiation in different regions meant that matter began to gather together into slightly denser clumps. These clumps provided the seeds from which galaxies later grew.

By the time the Universe had cooled to a temperature of 3000 K, about 300 000 years after the big bang, the mean energy of the photons had fallen to about 1 eV, and most of the matter in the Universe was in the form of neutral atoms. This was the trigger for another significant change in the behaviour of the Universe. The background radiation photons – those ten billion photons for every particle left over from the annihilation epoch (matter and antimatter reactions) – interacted for the last time with matter in the Universe. When hydrogen atoms are in their ground state, photons with an energy of at least 10 eV are required in order to excite them to even the next energy level. So from this point on in the history of the Universe, photons were no longer absorbed by matter. After this time, the cosmic background radiation simply expanded freely with the Universe, cooling as it did so.

When the cosmic microwave background radiation is observed today, very slight irregularities are observed in its temperature and intensity. These reflect slight differences in the matter distribution of the Universe at the time of the last interaction between the background photons and atoms. At the time of the discovery of these irregularities by the COBE satellite (Section 9.2), they were described as 'wrinkles in the fabric of space–time' (Figure 15.4a). The irregularities have since been mapped in even finer detail by the Wilkinson Microwave Anisotropy Probe (WMAP) satellite (Figure 15.4b), measurements from which have been claimed to herald the era of 'precision cosmology'.

Some time after the last interaction of matter and radiation, but before the Universe was a billion years old, the first galaxies formed. The exact time for this event is uncertain, but within these early galaxies, stars condensed out of

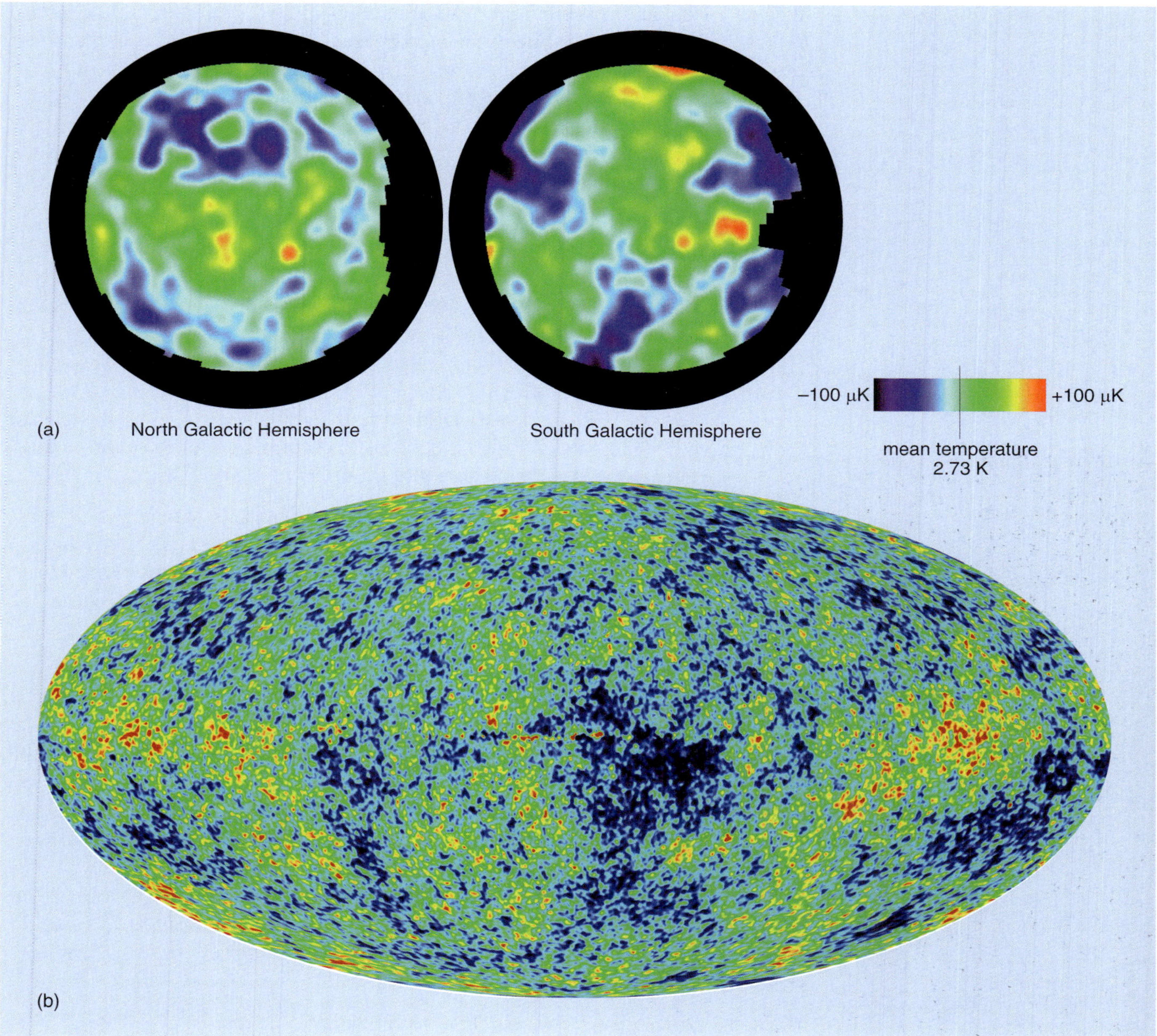

Figure 15.4 (a) This colour-coded map shows departures from uniformity in the cosmic microwave background radiation over the whole sky. The two panels correspond to two 'halves' of the sky, projected onto a flat picture. The scale represents the temperature either side of the mean temperature of 2.73 K. Violet regions are slightly cooler than the mean value of 2.73 K, red regions are slightly hotter, by about 100 μK (microkelvin). Most of the variations seen are believed to represent localised variations in the density of matter at a time 300 000 years after the big bang when this radiation interacted with matter for the last time. This map is the final result after four years operation of the COBE satellite. (b) This map, produced by the WMAP satellite, was released in 2003. It shows the whole sky. Ripples in the temperature of the microwave background are seen here on much finer scales than was possible with COBE.

the gas to become dense enough for nuclear reactions to start within their cores. Deep within these stars, hydrogen was converted into helium via nuclear fusion reactions, releasing energy as electromagnetic radiation into the Universe. As stars get older, so their cores contract and grow hotter, allowing helium fusion reactions to occur. These further reactions produce heavier nuclei such as carbon, oxygen and silicon. Low-mass stars, like the Sun, will end their cosmic chemistry here. They will eventually simply run out of nuclear fuel, and their cores will collapse to form dense, compact objects called **white dwarfs**.

The more massive the star though, the hotter is its interior, and the more massive the elements that can be produced by nuclear fusion reactions. But there is a limit to how far nuclear fusion can go. As noted in Section 6.6, when four protons are converted into a nucleus of helium-4, the products have a lower mass than the reactants. This mass difference is liberated as energy. Similar mass reductions apply for reactions to produce all the elements up to those with mass numbers in the range of about 56 to 62, such as iron, cobalt and nickel. However, for nuclear fusion reactions beyond this, more energy must be put into the reactions than is released from them, so these are not viable.

Figure 15.5 The Crab nebula, a supernova remnant in the constellation of Taurus. This expanding cloud of gas was thrown off in a supernova explosion when a massive star reached the end of its life. The cloud seen here is about 3 pc across. The exploding star was seen by Chinese astronomers on 4 July 1054, and was so bright that it remained visible in full daylight for 23 days.

As the most massive stars approach the ends of their lives, when their cores are composed of nuclei that undergo no more fusion, some nuclear reactions within their interiors release free neutrons. These neutrons can then add, slowly one at a time, to the iron, cobalt or nickel nuclei to make even more massive elements. As more and more neutrons are added, some transform into protons, via beta-minus decay, and in this way massive (stable) nuclei up to lead and bismuth can be created. This so called **s-process** (the 's' stands for slow) is the means by which the most massive, non-radioactive nuclei that exist in the Universe are created.

But what happens to these massive stars? When the core is largely composed of iron, they have no further source of energy available. The outer layers fall inwards, squeezing the centre of the star down until it has a density comparable to that of an atomic nucleus. The collapse halts – suddenly – and the material rebounds, setting off a shock wave back through the outer layers of the star. The result is a **supernova** (plural supernovae) explosion, in which 90% of the star's mass is thrown violently out into space (Figure 15.5). The star's core left behind will be revealed as a neutron star, or the ultimate compact object, a black hole.

In the final moments of its life, the star has one final surprise left. The immense temperatures and pressures created during the explosion cause electrons and protons to react to form huge numbers of free neutrons. These neutrons rapidly combine with existing nuclei to enable elements to be built *beyond* the lead and bismuth limit. In this **r-process** (the 'r' stands for rapid) during supernova explosions, all naturally radioactive elements in the Universe (apart from those that are the decay products of even more massive radioactive nuclei) were formed. A large proportion of the other nuclides between nickel and bismuth were also created in these violent events.

From here the star cycle repeats – but this time with a slight difference. Stars that formed after the first generation had lived and died had a richer source of raw material. A star like the Sun was formed in a galaxy that had already seen at least one generation of massive stars born, live and die in supernovae explosions. The gas and dust from which the Sun formed, about 5 billion years ago, had therefore been enriched by heavier elements produced inside these earlier stars. This leads to the possibility of the formation of planets from the material left behind.

The Earth itself formed from such debris. Every nucleus of carbon, oxygen, nitrogen and silicon found on the Earth and within living creatures was created inside the heart of an ancient star. Every nucleus of precious metal such as silver, gold and platinum was formed by slow neutron capture (s-process) in ageing stars, and every long-lived nucleus of a radioactive element such as thorium or uranium was formed by rapid neutron capture (r-process) during the supernovae explosions that mark stars' deaths. And so you have come full circle back to the present day, about 14 billion years after the big bang, when the Universe has cooled to only 2.73 K. A schematic summary of the entire history of the Universe from the big bang to the present day is shown in Figure 15.6.

Homing in on a fairly average spiral galaxy, you find a fairly average star somewhere out in one its spiral arms. Orbiting this star is a small rocky planet,

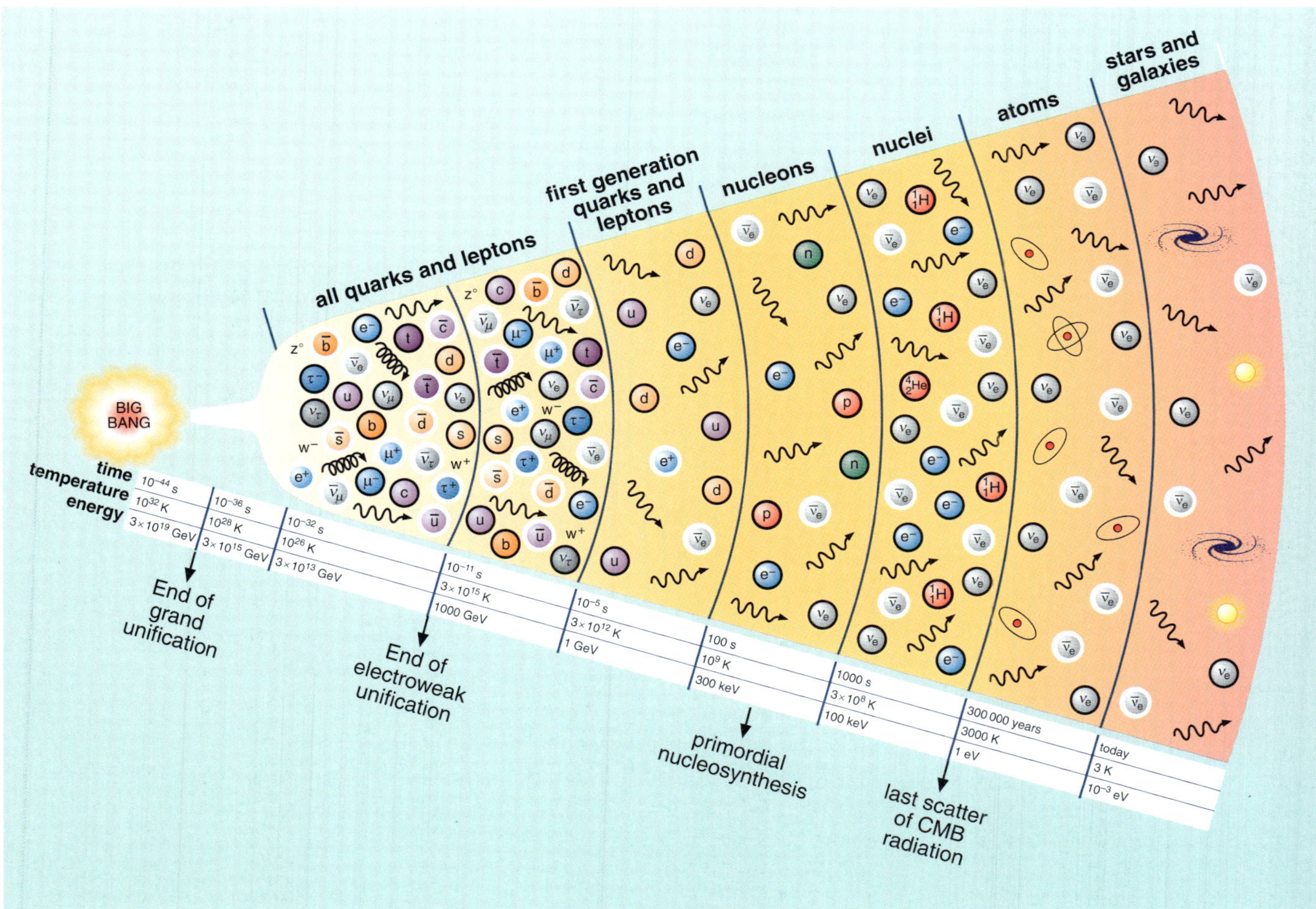

Figure 15.6 A schematic history of the Universe. See text for details.

two-thirds covered with water, and with an atmosphere rich in oxygen. On the surface of the planet are many living creatures, including members of one species who are so interested in the origin and complexity of the Universe that they build telescopes and particle accelerators with which to study it. They observe the expansion of the Universe by the redshift of distant galaxies, and the cooling of the Universe by the spectrum of its background radiation. Using particle accelerators they recreate extreme temperatures and examine particle reactions that have not occurred in the Universe for billions of years. The revelations of such experiments confirm that no epoch or location in the Universe is subject to any special dispensation, that at all times and all places the same physical principles hold, yet manifest themselves in a gloriously evolving diversity.

Question 15.1

Imagine a hypothetical universe in which weak interactions do not exist and in which only first-generation quarks and leptons are present (i.e. there are no charm, strange, top or bottom quarks, and no muons, muon neutrinos, tauons or tauon neutrinos). Speculate about the ways in which such a universe would be different from our own. Your answer should be between 100 and 150 words.

Question 15.2

(a) Describe three times or sites at which nucleosynthesis has occurred in the history of the Universe.

(b) At which of these times or sites did most of the (i) helium, (ii) oxygen, and (iii) uranium in the Universe originate?

Question 15.3

What are the three key pieces of observational evidence that support the idea of a hot big bang? Which of them do you think allows cosmologists to reach back furthest into the past, and why?

Activity 15.1 Universal history

We expect this activity will take you approximately 20 minutes.

(a) In which order did the following events occur in the history of the Universe? (*Hint*: consider the energy required for each process.)

- (i) the formation of atoms
- (ii) the formation of light nuclei
- (iii) the formation of quarks and leptons
- (iv) the formation of protons and neutrons
- (v) the annihilation of electrons and positrons
- (vi) the annihilation of quarks and antiquarks
- (vii) neutrinos cease to interact further with matter or radiation
- (viii) background photons cease to interact with matter

(b) Summarise the contents of the Universe at the times corresponding to the end of each of Sections 15.2–15.7.

Now look at the comments on this activity at the end of this book.

15.8 Summary of Chapter 15

The Universe was created at the instant of the big bang. As it has aged, the Universe has cooled and distances within it have increased. At the earliest times, the four fundamental interactions were unified, but as the temperature of the Universe decreased, so these interactions became distinct. The earliest time about which anything can be said is the Planck time, when the gravitational interaction had a similar strength to the other fundamental interactions. Before this, the concept of 'time' itself may have no meaning.

Early in its history, the Universe is presumed to have undergone an extremely rapid period of expansion, known as inflation. One effect of this was to smooth out any irregularities, leading to today's remarkably uniform observable Universe. The early Universe contained *almost* equal numbers of matter and antimatter particles (quarks and leptons). However, there was an asymmetry of a few parts in ten billion in favour of matter. The matter and antimatter underwent mutual annihilation and the result of this is that there are now about ten billion photons for every matter particle in the Universe.

Equal numbers of protons and neutrons were initially produced in the Universe from the up and down quarks remaining after annihilation. However, free neutrons decay, and this reduced their number, leading to a Universe containing about seven protons for every neutron today. All free neutrons were soon bound up within nuclei of deuterium, helium and lithium. The approximate distribution of mass in the Universe is about 25% helium-4 to 75% hydrogen, with small traces of other nuclei. Neutrinos ceased to interact with the rest of the Universe soon after protons and neutrons were formed.

At 300 000 years after the big bang, when the temperature was about 3000 K, photons produced from the matter–antimatter annihilations had their last interaction with the matter of the Universe. These photons, redshifted by a factor of a thousand by the expansion of the Universe, form the cosmic microwave background that is observed today. As the Universe cooled still further, galaxies and stars were able to form under the influence of gravity. Stars convert light nuclei into heavier ones within their cores. The more massive stars then undergo supernovae explosions, throwing material out into space ready to be included in later generations of stars and planets.

Chapter 16
Summary of Book 7

> What we have learned is like a handful of earth. What we have yet to learn is like the whole world.
>
> Avvaiyar, Indian poet-saint (9th century)

Congratulations on completing the scientific journey of exploration from quarks to quasars. In travelling this route, you will have gained an appreciation of how an understanding of the way the Universe behaves on small scales, and the rules that these smallest constituents obey, informs us about how the Universe behaves on the largest scales, and evolves with time. One of the underlying principles guiding the way on this journey has been the understanding that the *same* laws of physics apply at *all* times and in *all* places throughout the Universe. So, for instance, the principle of the conservation of energy and the behaviour of light that you read about in Book 3 are still valid when considering the behaviour of everything from quarks to quasars.

The fundamental constituents of matter have been revealed as particles called leptons and quarks. In particular, the world around us today comprises only first-generation leptons (electrons and electron neutrinos) and first-generation quarks (up and down) which are joined together in triplets called protons (uud) and neutrons (udd), that in turn make up nuclei and atoms. Each particle has an associated antimatter particle with opposite electric charge and colour charge (where appropriate). These fundamental particles interact with each other by way of four fundamental interactions (known as the electromagnetic, strong, weak and gravitational interactions) via the exchange of other particles (known as photons, gluons, W and Z bosons, and gravitons).

One of the cornerstones of science is quantum physics, which tells us that the smallest constituents of matter, such as nuclei, atoms and molecules, can only exist with specific (quantised) values of energy, known as energy levels. It is the jumps between these energy levels that give rise to the emission and absorption of photons, which we perceive as electromagnetic radiation. The other important feature of quantum physics is that the smallest particles of matter have indeterminate positions and velocities. Bearing this in mind, the 'best' model of atoms envisages electrons as existing in fuzzy clouds around the central nucleus. Different quantum states of an atom, described by a set of four quantum numbers, correspond to different energy levels and different electron clouds.

The four fundamental interactions are best expressed in terms of quantum theories: quantum electrodynamics describes both electromagnetic and weak interactions, whilst quantum chromodynamics describes strong interactions. A quantum theory of gravity, however, has yet to be formulated. Because of this, the approximations to reality described by Newton's law of gravity are used to explain most everyday gravitational interactions, and Einstein's general theory of relativity is used to explain gravitational interactions when the masses and accelerations become very large. Vital insight into the way the Universe evolves comes from the understanding that the strength of the four fundamental interactions varies with the energy of interaction. At the highest energies, the four interactions are believed to exist in a single superunified form. As the energy decreases, so the gravitational, strong, weak and electromagnetic interactions in turn become separated into the four distinct interactions perceived in the Universe today.

This breakdown of unification as energy decreases is the key to understanding the evolution of the Universe. Observations of distant galaxies reveal that their spectra are all redshifted, in accord with the Hubble relationship, implying that the Universe is expanding. Observations of the cosmic microwave background radiation imply that the Universe is cooling. These two observations together tell us that the Universe began in a hot big bang, at which instant, time and space were created. The Universe has cooled and expanded ever since. Particle reactions and decays occurring in the first few minutes of the Universe's history set the scene for the structures observed in the Universe today. So now you have come a full circle from quarks to quasars and Figure 16.1, in a sense, summarises the scientific journey that you have travelled in this book.

However, you have explored more than just a range of scientific facts and theories as you've studied this book. You've also practised and developed a variety of mathematical skills, including dealing with very large and very small numbers in scientific notation, and combining algebraic quantities with different units. As well as this, you have learned how to cope with a range of unfamiliar

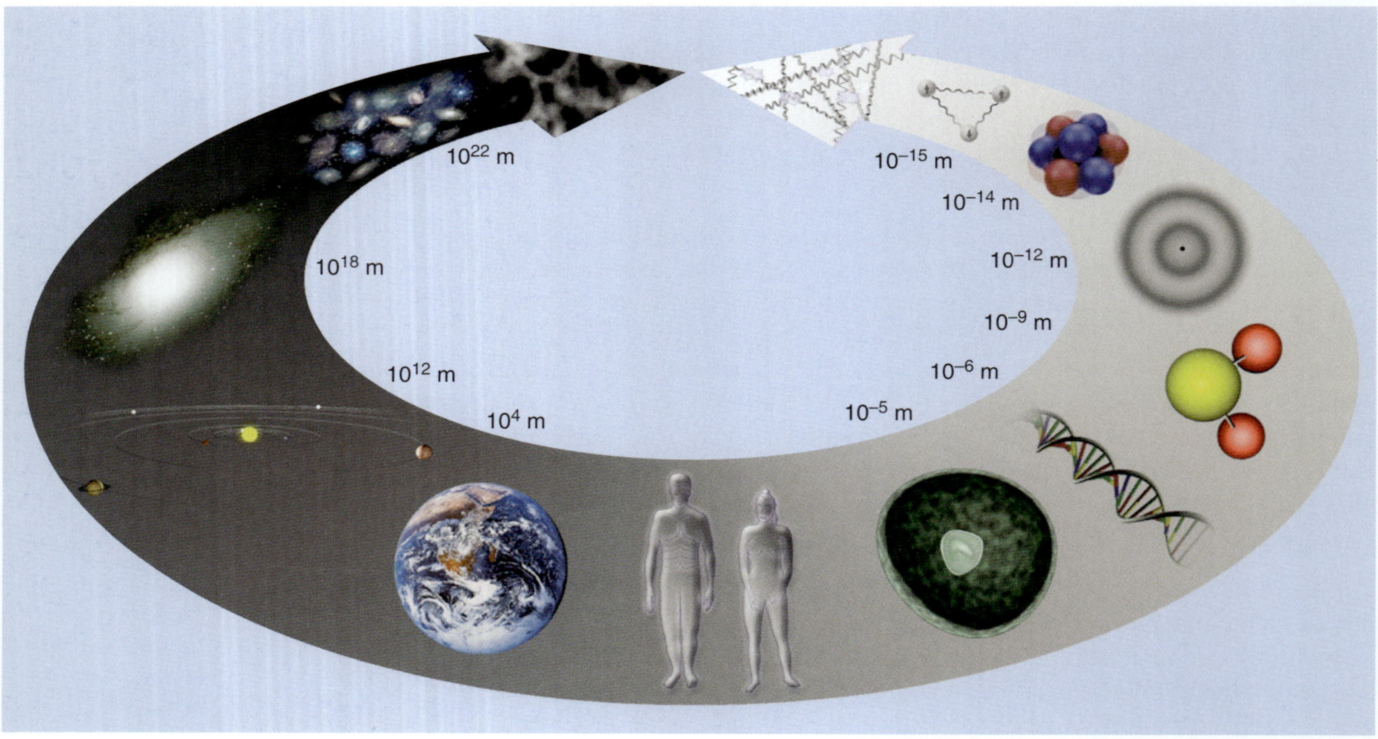

Figure 16.1 People stand midway between the immensity of the Universe and the unimaginable smallness of atomic nuclei and their building blocks. Yet what happened long ago involving these tiniest objects set the entire Universe on its course. Today, we cannot understand the Universe as a whole without understanding nature on the smallest scales. That is why the largest and smallest meet in this picture.

concepts and practised applying what you have learned to many different problems.

In the next, and final, book of the course, you will be able to use the skills and knowledge that you have learned here and in earlier books to explore one of the most fascinating topics in science today – life in the Universe. All the areas of science from astronomy to biology, chemistry, geology and physics are needed to appreciate questions such as how life evolved on Earth, where we might find life in the Solar System or elsewhere in the Galaxy, and how scientists are now finding planets around other stars and searching for evidence of life in the Universe.

Answers to questions

$E_{ph} = E_x - E_1$

$E_x = E_{ph} + E_1$

$=$

Question 3.1

(a) The energy of the photon is given by the (positive) difference between the two energies, in this case $E_4 - E_3$. Since the final energy (E_4) is higher than the initial (E_3) in this case, the photon would have to be absorbed.

(b) In this case the final energy is less than the initial energy, so a photon of energy $E_2 - E_1$ is emitted.

(c) Figure 3.4a shows that the transition from E_7 to E_1 corresponds to an energy difference that is substantially more than that involved in the transition from E_7 to E_2. But the transition between E_7 and E_2 corresponds to an energy difference of 3.12 eV which is near the *upper* energy limit of the visible spectrum. The photon associated with the E_7 to E_1 transition is therefore not in the visible part of the spectrum. [In fact this transition corresponds to a photon of ultraviolet radiation.]

(d) Figure 3.4a shows that the transition from E_5 to E_6 corresponds to an energy difference that is substantially less than that involved in the transition from E_2 to E_3, for instance. But the transition between E_2 and E_3 corresponds to an energy difference of 1.89 eV which is near the *lower* energy limit of the visible spectrum. The photon associated with the E_5 to E_6 transition is therefore not in the visible part of the spectrum. [In fact this transition corresponds to a photon of infrared radiation.]

Question 3.2

(a) The lines of the Lyman series all correspond to transitions down to the $n = 1$ energy level of hydrogen. Using the values for the energy levels from Figure 3.5, the first five lines will have the following energies:

$$E_2 - E_1 = (-3.40\ \text{eV}) - (-13.60\ \text{eV}) = 10.20\ \text{eV}$$

$$E_3 - E_1 = (-1.51\ \text{eV}) - (-13.60\ \text{eV}) = 12.09\ \text{eV}$$

$$E_4 - E_1 = (-0.85\ \text{eV}) - (-13.60\ \text{eV}) = 12.75\ \text{eV}$$

$$E_5 - E_1 = (-0.54\ \text{eV}) - (-13.60\ \text{eV}) = 13.06\ \text{eV}$$

$$E_6 - E_1 = (-0.38\ \text{eV}) - (-13.60\ \text{eV}) = 13.22\ \text{eV}$$

(b) The highest-energy photons corresponding to the Lyman series will have energy:

$$E_\infty - E_1 = (0\ \text{eV}) - (-13.60\ \text{eV}) = 13.60\ \text{eV}$$

(c) Since these energies are all greater than 3.2 eV (the upper limit for visible radiation), these lines will all lie in the ultraviolet part of the electromagnetic spectrum.

(d) If they were equally 'bright' the lines of the Lyman series would form a 'bunched' pattern of the kind shown in Figure 3.10.

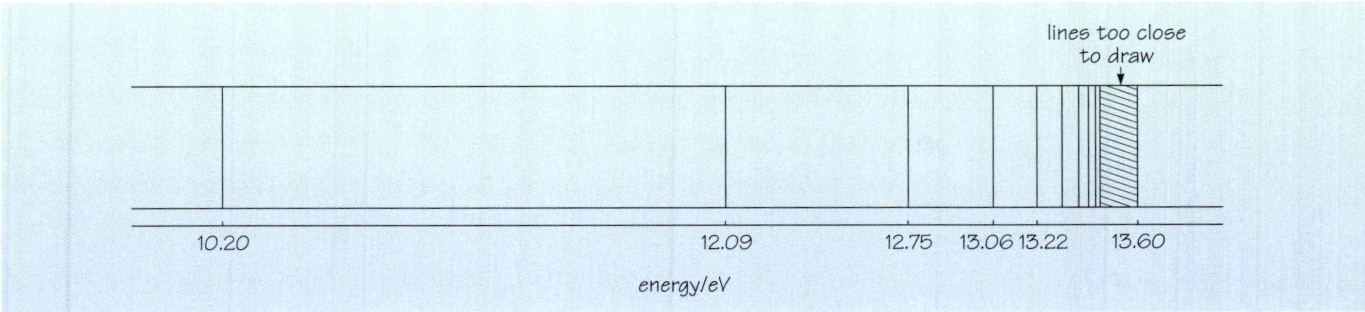

Figure 3.10 Answer to Question 3.2. The Lyman series of the hydrogen spectrum.

Question 3.3

For $n = 3$, the allowed values of l are $l = 0$, $l = 1$, and $l = 2$. When $l = 0$, then $m_l = 0$ and $m_s = +\frac{1}{2}$ or $-\frac{1}{2}$; when $l = 1$, then $m_l = -1$, 0 or +1 and in each case m_s may be $+\frac{1}{2}$ or $-\frac{1}{2}$; finally when $l = 2$, then $m_l = -2$, -1, 0, +1 or +2 and in each case m_s may be $+\frac{1}{2}$ or $-\frac{1}{2}$. Hence there are $2 \times ((2 \times 0) + 1) = 2$ states with $l = 0$, there are $2 \times ((2 \times 1) + 1) = 6$ states with $l = 1$, and $2 \times ((2 \times 2) + 1) = 10$ states with $l = 2$. The total number of possibilities is, therefore, $2 + 6 + 10 = 18$, and this is equal to $2n^2$.

Question 3.4

(a) A 1s quantum state has $n = 1$ and $l = 0$. So the third quantum number m_l must also be equal to 0 according to the rules. The fourth quantum number m_s can be either $+\frac{1}{2}$ or $-\frac{1}{2}$. So the two possible sets of quantum numbers (n, l, m_l, m_s) that the atom can have are $(1, 0, 0, +\frac{1}{2})$ and $(1, 0, 0, -\frac{1}{2})$.

(b) Combining Equations 2.3 and 2.4, the photon energy corresponding to a given wavelength of electromagnetic radiation is:

$$E_{\text{ph}} = \frac{hc}{\lambda}$$

so, in this case:

$$E_{\text{ph}} = \frac{(4.1 \times 10^{-15}\ \text{eV s}) \times (3.0 \times 10^{8}\ \text{m s}^{-1})}{0.21\ \text{m}} = 5.9 \times 10^{-6}\ \text{eV}$$

Hence when a hydrogen atom makes a transition between the two quantum states corresponding to the sets of quantum numbers in (a) it emits a photon whose energy is 5.9 μeV (microelectronvolts). This tiny energy corresponds to the separation of the two energy levels that comprise the 1s quantum state. (It emphasises why we are justified in making the approximation that *all* quantum states with the same principal quantum number essentially have the same energy.)

Question 3.5

(a) The spectrum produced by a glowing copper wire would be a continuous spectrum. The spectrum would contain light and infrared radiation, spread over a whole range of energies.

(b) The spectrum produced by a vapour of copper atoms that are excited by an electric current would be an emission spectrum. The spectrum would contain light only at certain discrete energies; that is, it would contain spectral lines.

Question 4.1

(a) If the hydrogen atom is isolated from other sources of energy, then it cannot jump from its ground state to an excited state. Also, there is no lower energy level to which the atom could jump, so the statement is correct. (However, ensuring that the atom is indeed 'isolated from other sources of energy' is harder than it sounds. As you will see later in the book, energy may be 'borrowed' for a short time from 'empty space', as long as the energy debt is soon paid back.)

(b) From Equation 3.1, the second energy level of hydrogen has an energy of:

$$E_2 = \frac{-13.60\ \text{eV}}{2^2} = -3.40\ \text{eV}$$

So a photon energy of (–13.60 eV) – (–3.40 eV) = 10.20 eV is needed to excite the atom to the second energy level. If each photon has less than 10 eV of energy there is no way any of them can excite a hydrogen atom from its ground state, so this statement too is correct.

(c) If a hydrogen atom in its ground state absorbs a photon of energy 12.75 eV it will make a transition to an excited state with an energy of (13.60 eV) + (12.75 eV) = –0.85 eV. This is the E_4 energy level, since:

$$E_4 = \frac{-13.60\ \text{eV}}{4^2} = -0.85\ \text{eV}$$

From here the atom can make transitions to *any* of the lower energy levels. Only those transitions ending up at the E_2 energy level will involve emission of visible photons. Other transitions, to the E_3 or E_1 levels, will result in the emission of infrared or ultraviolet photons respectively. So although the atom *could* emit a visible photon, we cannot say that it definitely *will* do so. Once again, therefore, this statement is correct.

Question 4.2

(a) This is not correct as it is too vague. The Heisenberg uncertainty principle says that if you know the position of a particle precisely, then there is a large uncertainty in its velocity, but there is nothing in the principle that prevents a precise measurement of velocity.

(b) This is correct (it is mentioned in the text).

(c) This is not correct. The Heisenberg uncertainty principle asserts that the indeterminacies are unavoidable as they are inherent in the nature of the Universe. The existence of such limitations on our knowledge is just 'the way things are'.

(d) This is certainly not correct. Because the limitations of his principle are an unavoidable aspect of the nature of the Universe, experimenters cannot possibly be expected to overcome them. All that can be asked of experimenters is to minimise the uncertainties of their measurements within the limits set by the uncertainty principle.

Question 5.1

(a) For a Be^{3+} ion, the nuclear charge $Z = 4$, so the ground-state energy may be calculated from Equation 5.1 as:

$$E_1 = 4^2 \times \left(\frac{-13.60\ \text{eV}}{1^2}\right) = 16 \times (-13.60\ \text{eV}) = 217.6\ \text{eV}$$

(b) The energy levels of Be^{3+} are like those of H, He^+ and Li^{2+}, all of which contain a single bound electron. The energy of the first excited state of the Be^{3+} ion may be found by setting $Z = 4$ and $n = 2$ in Equation 5.1. In this case:

$$E_2 = 4^2 \times \frac{-13.60\ \text{eV}}{2^2} = -54.40\ \text{eV}$$

This is the same result as with $Z = 2$ and $n = 1$, corresponding to the ground state of the He^+ ion, namely:

$$E_1 = 2^2 \times \frac{-13.60\ \text{eV}}{1^2} = -54.40\ \text{eV}$$

Question 5.2

(a) Since the energy levels are $Z^2 = 4$ times further apart, the corresponding jump produces a photon with four times more energy, i.e. $4 \times 1.89\ \text{eV} = 7.56\ \text{eV}$. [This is in the ultraviolet part of the electromagnetic spectrum.]

Alternatively, this can be checked using Equation 5.1. The $n = 3$ and $n = 2$ energy levels for He^+ with $Z = 2$ may be calculated as:

$$E_3 = 2^2 \times \frac{-13.60\ \text{eV}}{3^2} = -6.04\ \text{eV}$$

and

$$E_2 = 2^2 \times \frac{-13.60\ \text{eV}}{2^2} = -13.60\ \text{eV}$$

So the photon energy emitted is given by:

$$E_3 - E_2 = (-6.04\ \text{eV}) - (-13.60\ \text{eV}) = (13.60 - 6.04)\ \text{eV} = 7.56\ \text{eV}$$

which is four times larger than the energy of the photons emitted by hydrogen atoms making a similar transition (i.e. 1.89 eV as shown in Figure 3.4a).

(b) In Equation 5.1, Z^2 appears on the top line whilst n^2 appears on the bottom line. Now, in the He^+ ion, Z is twice what it is in the hydrogen atom, so to get the same photon energy from a transition, the values of n must also be doubled. The jump from $n = 6$ to $n = 4$ in the helium ion therefore gives the same photon energy as that from $n = 3$ to $n = 2$ in hydrogen.

To check this, the $n = 6$ and $n = 4$ levels in He^+ are given by:

$$E_6 = 2^2 \times \left(\frac{-13.60\text{ eV}}{6^2}\right) = -1.51\text{ eV}$$

and

$$E_4 = 2^2 \times \left(\frac{-13.60\text{ eV}}{4^2}\right) = -3.40\text{ eV}$$

So the photon energy emitted is given by:

$$E_6 - E_4 = (-1.51\text{ eV}) - (-3.40\text{ eV}) = (3.40 - 1.51)\text{ eV} = 1.89\text{ eV}$$

which is the same as $E_3 - E_2$ for hydrogen.

Question 5.3

(a) Setting $Z = 3$ in Equation 5.3:

$$Z_{\text{est}} = (3 - 0.3125) = 2.6875$$

So, using Equation 5.2, an estimate for the ground-state energy of Li^+ is:

$$E_{\text{est}} = 2 \times (2.6875)^2 \times (-13.60\text{ eV}) = -196.5\text{ eV}$$

(b) Setting $Z = 3$ and $n = 1$ in Equation 5.1:

$$E_1 = 3^2 \times \left(\frac{-13.60\text{ eV}}{1^2}\right) = -122.4\text{ eV}$$

for the ground-state energy of Li^{2+}.

(c) The energy that must be supplied to a Li^+ ion in order to remove one more electron and turn it into a Li^{2+} ion is therefore:

$$E_{\text{ionisation}} = (-122.4\text{ eV}) - (-196.5\text{ eV}) = 74.1\text{ eV}.$$

So the estimate is that about 74.1 eV must be supplied to the Li^+ ion in order to ionise it further to Li^{2+}. This is illustrated in Figure 5.3.

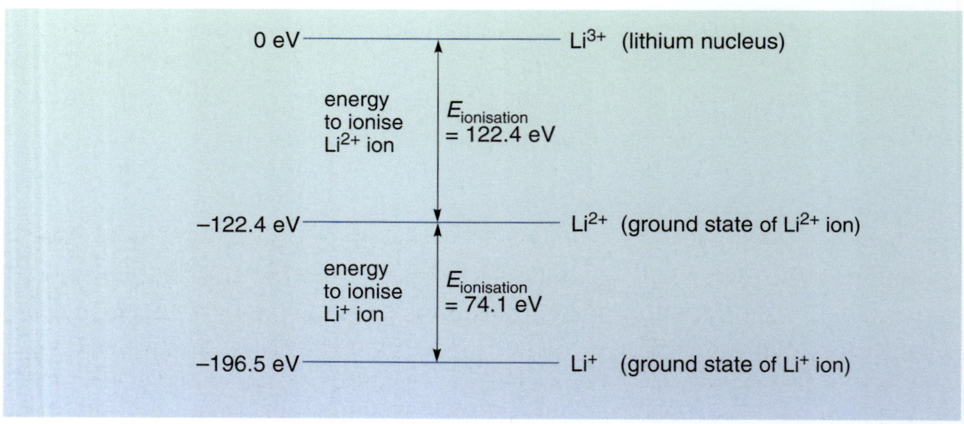

Figure 5.3 Part of the energy-level diagram for lithium ions.

Question 5.4

(a) Since each 's' state can accommodate two electrons, the ground-state electron configuration for beryllium is $1s^2 2s^2$.

(b) Since the fifth electron cannot go into either the 1s state or the 2s state, which are filled with four electrons, the ground-state electron configuration for boron is $1s^2 2s^2 2p^1$.

(c) The 1s, 2s and 2p subshells can accommodate a total of $2 + 2 + 6 = 10$ electrons. [This corresponds to the ground state of the neon atom, $1s^2 2s^2 2p^6$.]

Question 6.1

(a) The change in binding energy = (−1731.6 MeV) − (−1708.2 MeV − 28.3 MeV) = 4.9 MeV.

Therefore the energy liberated by the decay is 4.9 MeV.

(b) The change in binding energy = (−1759.9 MeV) − (−1736.7 MeV − 28.3 MeV) = 5.1 MeV. Since 5.1 MeV of energy is liberated by the decay, the mass of the products must be 5.1 MeV/c^2 less than that of the original nucleus.

Question 6.2

The nitrogen isotope will undergo beta-minus decay as follows:

$$^{16}_{7}\text{N} \longrightarrow {}^{16}_{8}\text{O} + \text{e}^- + \overline{\nu}_\text{e}$$

Since the resulting nucleus has an atomic number of eight, this is an isotope of oxygen. [Note that the atomic number Z *increases* by one, and the mass number A remains unchanged, and this is true for *all* beta-minus decays, whatever the initial nucleus.]

Question 6.3

The phosphorus isotope will undergo beta-plus decay as follows:

$$^{30}_{15}\text{P} \longrightarrow {}^{30}_{14}\text{Si} + \text{e}^+ + \nu_\text{e}$$

Since the resulting nucleus has an atomic number of fourteen, this is an isotope of silicon. [Note that the atomic number Z *decreases* by one, and the mass number A remains unchanged, and this is true for *all* beta-plus decays, whatever the initial nucleus.]

Question 6.4

Beta-minus and beta-plus decay are clearly rather similar processes. In the first, a neutron transforms into a proton with the emission of an electron and an electron antineutrino; in the second, a proton transforms into a neutron with the emission of an antielectron and an electron neutrino. So two particles are emitted in each case; one is a matter particle and the other is an antimatter particle. They differ though because, in beta-minus decay, the atomic number *increases* by one, and in beta-plus decay it *decreases* by one. In both types of decay, the mass number of the nucleus remains unchanged.

In the electron capture process, a proton is transformed into a neutron, as in the case of beta-plus decay. Here too the atomic number *decreases* by one and the mass number is unchanged. However, in this case, a matter particle is captured (an electron) and another matter particle is emitted (an electron neutrino).

Question 6.5

The mass number of the reactants is 235 + 1 = 236. The strontium-99 nucleus and the two free neutrons released account for a mass number of 99 + 2 = 101. So the other nucleus produced must have a mass number of 236 – 101 = 135. The atomic number of the original uranium nucleus is 92, whilst that of the strontium nucleus is 38. So the atomic number of the other nucleus must be 92 – 38 = 54.

In fact the other nucleus produced is xenon-135, and the reaction may be represented as:

$$^{235}_{92}\mathrm{U} + ^{1}_{0}\mathrm{n} \longrightarrow ^{236}_{92}\mathrm{U} \longrightarrow ^{99}_{38}\mathrm{Sr} + ^{135}_{54}\mathrm{Xe} + 2^{1}_{0}\mathrm{n}$$

Question 6.6

(a) The fusion process may be written as:

$$^{4}_{2}\mathrm{He} + ^{4}_{2}\mathrm{He} + ^{4}_{2}\mathrm{He} \longrightarrow ^{12}_{6}\mathrm{C}$$

(b) The binding energy of each helium-4 nucleus is –28.3 MeV, whilst that of the carbon-12 nucleus is –92.2 MeV. The energy released by the triple-alpha process is therefore (–28.3 MeV × 3) – (–92.2 MeV) = 7.3 MeV.

Question 7.1

(a) You have seen that the quark content of a proton is (uud), so the antiquark content of an antiproton must be $(\overline{\mathrm{u}}\ \overline{\mathrm{u}}\ \overline{\mathrm{d}})$. Now, the charge of an antiquark is opposite to that of the corresponding quark. So the charge of a $\overline{\mathrm{u}}$ antiquark is $-\frac{2}{3}e$ whilst the charge of a $\overline{\mathrm{d}}$ antiquark is $+\frac{1}{3}e$. The charge of an antiproton is therefore $-\frac{2}{3}e - \frac{2}{3}e + \frac{1}{3}e = -e$. This is the opposite charge to that of a proton, as expected.

(b) Similarly, you have seen that the quark content of a neutron is (udd), so the antiquark content of an antineutron must be $(\overline{\mathrm{u}}\ \overline{\mathrm{d}}\ \overline{\mathrm{d}})$. Using the values for the charge of a $\overline{\mathrm{u}}$ antiquark and a $\overline{\mathrm{d}}$ antiquark from above, the charge of an antineutron is therefore $-\frac{2}{3}e + \frac{1}{3}e + \frac{1}{3}e = 0$. Notice that this is the *same* as the charge of a neutron, even though the antineutron is composed of antiquarks.

Question 7.2

Pions and nucleons contain only up and down quarks, so these are the only 'raw materials' available from which to build the new hadron. To get a charge of +2*e* requires three up quarks $(+\frac{2}{3}e + \frac{2}{3}e + \frac{2}{3}e = +2e)$. This hadron can be formed by the collision of π^+, $(\mathrm{u}\overline{\mathrm{d}})$, with a proton, (uud), followed by the annihilation of $\overline{\mathrm{d}}$ with d. Since the hadron contains three quarks, it is a baryon.

Question 7.3

When two neutrons collide with a combined kinetic energy of 500 MeV, this is enough energy to make three pions (with mass energy 140 MeV each). In addition, all the possible reactions to make either one or two pions can also occur. So the total range of possibilities is as follows:

$$n + n \longrightarrow n + n + \pi^0$$

$$n + n \longrightarrow n + p + \pi^-$$

$$n + n \longrightarrow n + n + \pi^0 + \pi^0$$

$$n + n \longrightarrow n + n + \pi^+ + \pi^-$$

$$n + n \longrightarrow n + p + \pi^- + \pi^0$$

$$n + n \longrightarrow p + p + \pi^- + \pi^-$$

$$n + n \longrightarrow n + n + \pi^0 + \pi^0 + \pi^0$$

$$n + n \longrightarrow n + n + \pi^+ + \pi^- + \pi^0$$

$$n + n \longrightarrow n + p + \pi^- + \pi^0 + \pi^0$$

$$n + n \longrightarrow n + p + \pi^+ + \pi^- + \pi^-$$

$$n + n \longrightarrow p + p + \pi^- + \pi^- + \pi^0$$

Question 8.1

(a) Since 1.0 pc = 3.1×10^{13} km, a distance of 200 Mpc is equivalent to about $200 \times 10^6 \times 3.1 \times 10^{13}$ km = 6.2×10^{21} km.

(b) Since 1.0 pc = 3.3 ly, a distance of 200 Mpc is equivalent to $200 \times 10^6 \times 3.3 = 6.6 \times 10^8$ light-years. So a beam of light would take 660 million years to travel from the cluster of galaxies to Earth.

Question 8.2

(a) Galaxy G1 is $\dfrac{2250\ \text{kpc}}{250\ \text{kpc}} = 9$ times further away than galaxy G2. As they have the same luminosity, G1 will therefore appear to be $9^2 = 81$ times fainter than G2.

(b) Galaxy G3 is 36 times fainter than galaxy G4. As they have the same luminosity, G3 must be six times further away than G4 (since $6^2 = 36$). Since G3 is 1500 kpc away, G4 must be only $\dfrac{1500\ \text{kpc}}{6} = 250$ kpc away.

Question 8.3

(a) Rearranging Equation 8.3:

$$z = \frac{v}{c} = \frac{0.60\ \text{km s}^{-1}}{3.0 \times 10^5\ \text{km s}^{-1}} = 2.0 \times 10^{-6}$$

[Redshift is a pure number and so has no units.]

(b) The shift in wavelength is the difference between the observed wavelength and the rest wavelength. This may be determined by rearranging Equation 8.2 to give $\Delta\lambda = z\lambda_0$. So in this case:

$$\Delta\lambda = (2.0 \times 10^{-6} \times 656\ \text{nm}) = 1.3 \times 10^{-3}\ \text{nm}.$$

This is a tiny shift and for this reason the Doppler effect with light is not normally noticeable in everyday situations.

Question 8.4

(a) Since the wavelength of the light from the galaxy is shifted towards longer wavelengths, i.e. towards the red, the galaxy must be receding from the Earth.

(b) Using Equation 8.2:

$$z = \frac{\Delta\lambda}{\lambda_0} \text{ so, in this case, the redshift, } z = \frac{(500.7 - 486.1)\ \text{nm}}{486.1\ \text{nm}} = 0.0300$$

(to three significant figures)

(c) Equation 8.3 states that the apparent speed of the galaxy, $v = zc$, so here $v = 0.0300 \times 3.0 \times 10^5\ \text{km s}^{-1} = 9.0 \times 10^3\ \text{km s}^{-1}$ (to two significant figures).

Question 8.5

(a) The first step is to calculate the apparent recession speed of the galaxy using Equation 8.4, $v = H_0 r$. Taking the Hubble constant as 70 km s^{-1} Mpc^{-1} and the distance r as 300 Mpc, the apparent speed of the galaxy is:

$$(70\ \text{km s}^{-1}\ \text{Mpc}^{-1}) \times (300\ \text{Mpc}) = 2.1 \times 10^4\ \text{km s}^{-1}.$$

The redshift of this galaxy can be found by rearranging Equation 8.4 to give:

$$z = \frac{v}{c}$$

so in this case, the redshift $= \dfrac{2.1 \times 10^4\ \text{km s}^{-1}}{3.0 \times 10^5\ \text{km s}^{-1}} = 0.070.$

(b) The first step here is to calculate the apparent recession speed of the galaxy using Equation 8.3, $v = zc$, so in this case the apparent speed of the galaxy is $0.056 \times 3.0 \times 10^5\ \text{km s}^{-1} = 1.68 \times 10^4\ \text{km s}^{-1}$. Then rearranging Equation 8.5 to give:

$$r = \frac{v}{H_0}$$

the distance may be calculated as:

$$r = \frac{1.68 \times 10^4\ \text{km s}^{-1}}{70\ \text{km s}^{-1}\ \text{Mpc}^{-1}} = 240\ \text{Mpc}$$

Question 9.1

(a) 230 °C is equivalent to about 500 K (since 273 + 230 is about 500). From Figure 9.1 it can be estimated that an object at a temperature of 500 K would radiate a black-body spectrum whose peak intensity occurs in the infrared region of the electromagnetic spectrum.

(b) From Figure 9.1 it may be estimated that an object whose black-body spectrum has a peak intensity in the ultraviolet part of the electromagnetic spectrum must be at a temperature of between about 10^4 K and 10^6 K.

Question 10.1

(a) Both nuclei are positively charged, so the electromagnetic force between them is repulsive.

(b) The oxygen-16 nucleus has eight protons, whereas the nitrogen-16 nucleus only had seven protons. The electromagnetic force depends on the product of the two charges, so the repulsive electromagnetic force will increase by a factor of 8/7 (i.e. about 1.14 times its previous value or by 14%).

(c) If both nuclei were converted into neutral atoms, by the addition of eight electrons to each nucleus, the electromagnetic force would reduce to zero.

Question 10.2

(a) Using Equation 10.2, the size of the force F_e exerted on a charge $Q = +2e$ when it is placed in an electric field of strength $\mathscr{E} = 200\ \mathrm{N\ C^{-1}}$ is given by:

$$F_e = Q\mathscr{E} = (2 \times 1.6 \times 10^{-19}\ \mathrm{C}) \times 200\ \mathrm{N\ C^{-1}}$$

$$F_e = 6.4 \times 10^{-17}\ \mathrm{N}$$

(b) The charge on a particle in an electric field, which experiences a certain electric force, is given by a rearrangement of Equation 10.2 as:

$$Q = \frac{F_e}{\mathscr{E}}$$

In this case therefore:

$$\text{charge, } Q = \frac{3.2 \times 10^{-12}\ \mathrm{N}}{5.0 \times 10^{6}\ \mathrm{N\ C^{-1}}} = 6.4 \times 10^{-19}\ \mathrm{C}$$

Given that the charge of a proton, $e = 1.6 \times 10^{-19}$ C, this is equivalent to a charge of $+4e$.

Question 10.3

Before the work of Maxwell, Coulomb had shown that stationary electric charges give rise to electric fields, whilst Oersted and Ampère had shown that moving electric charges (currents) give rise to magnetic fields. Faraday and Henry had also shown that a changing magnetic field produces an electric field. Maxwell's predictions were that a changing electric field gives rise to a magnetic field and that, as a result, electromagnetic waves must exist that travel at the speed of light.

Question 10.4

(a) An energy of 1.0 MeV ($= 1.0 \times 10^6$ eV) implies that the maximum time by which the energy debt must be repaid is given by:

$$\Delta t = \frac{3.3 \times 10^{-16}\ \mathrm{eV\ s}}{\Delta E} = \frac{3.3 \times 10^{-16}\ \mathrm{eV\ s}}{1.0 \times 10^{6}\ \mathrm{eV}} = 3.3 \times 10^{-22}\ \mathrm{s}$$

[This is an incredibly short timescale.]

(b) The energy debt could be repaid by the electron and positron annihilating each other.

Question 11.1

A graph summarising how the strengths of the electromagnetic and strong interactions vary between 100 GeV and 10^{15} GeV is shown in Figure 11.6.

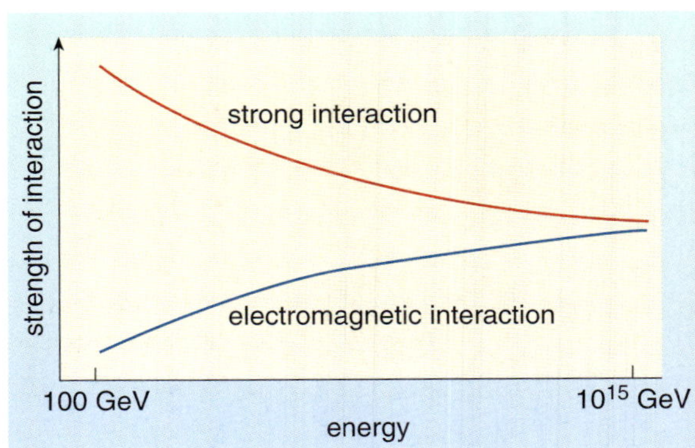

Figure 11.6 Answer to Question 11.1.

Question 11.2

Table 11.2 is the completed version of Table 11.1. Only particles with non-zero electric charge interact with photons. Only particles with colour charge interact with gluons.

Table 11.2 Completed version of Table 11.1.

Particle	Electric charge	Colour charge	Quanta with which the particle interacts
electron	$-e$	–	photons
electron neutrino	0	–	–
up quark	$+\frac{2}{3}e$	red, green or blue	photons, gluons
down quark	$-\frac{1}{3}e$	red, green or blue	photons, gluons
photon	0	–	–
gluon	0	colour and anticolour	gluons

Question 12.1

As shown by the diagram in Figure 12.3, beta-plus decay involves an up quark (u) changing into a down quark (d), where a W^+ boson is created. Shortly thereafter, the energy account is balanced by the production of a positron (e^+) and an electron neutrino (ν_e).

The electric charge initially is that of an up quark, i.e. $+\frac{2}{3}e$. The products of the decay are a down quark (electric charge, $-\frac{1}{3}e$), a positron (electric charge, $+e$) and an electron neutrino (electric charge, 0). The net electric charge after the decay is therefore $-\frac{1}{3}e+e=+\frac{2}{3}e$, the same as it was initially. Electric charge is therefore conserved.

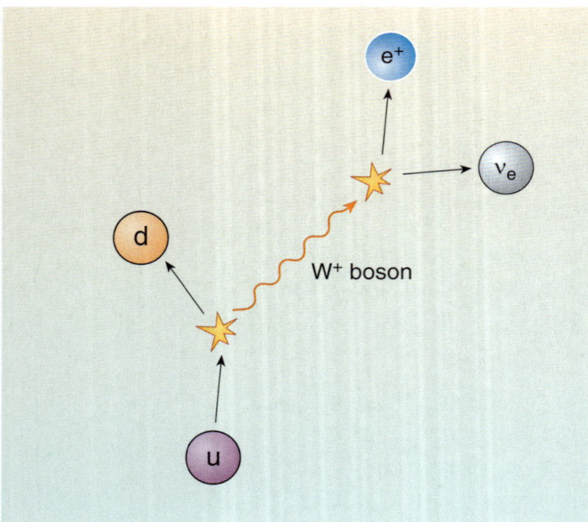

Figure 12.3 A diagram showing a beta-plus decay process in which an up quark converts into a down quark.

There is one quark present both before and after the decay, so the total number of quarks minus the number of antiquarks is conserved, and equal to one. There are no leptons present initially, but one lepton (the electron neutrino) and one antilepton (the positron) present at the end. Therefore the total number of leptons minus the number of antileptons is also conserved, and equal to zero.

Question 12.2

The strength of the electromagnetic interaction *increases* with increasing energy and the strength of the strong interaction *decreases* with increasing energy. At 100 GeV the strong interactions is ten times stronger than the electromagnetic interaction; whilst at 10^{15} GeV their strengths are predicted to be equal.

The strength of the weak interaction *increases* with increasing energy. At 100 GeV its strength is comparable to that of the electromagnetic interaction, whereas at an energy of 1 GeV, it is a factor of 10^8 smaller.

Question 13.1

Equations 13.1 and 13.2 both give expressions for the gravitational force of attraction. They may be equated, providing that m_2 and r in Equation 8.1 are identified with the mass and radius of the Earth. So, setting Equations 13.1 and 13.2 equal to each other gives:

$$G \times \frac{m_1 m_2}{r^2} = m_1 g$$

The mass of the object m_1 cancels from each side, leaving:

$$G \times \frac{m_2}{r^2} = g$$

This may be rearranged to make m_2, the mass of the Earth, the subject of the equation:

$$m_2 = \frac{g r^2}{G}$$

Putting in the numbers, this becomes:

$$m_2 = \frac{(9.8 \text{ m s}^{-2}) \times (6.4 \times 10^6 \text{ m})^2}{6.7 \times 10^{-11} \text{ N m}^2 \text{ kg}^{-2}}$$

Since the unit of newtons is equivalent to kg m s^{-2}:

$$m_2 = \frac{(9.8 \text{ m s}^{-2}) \times (4.1 \times 10^{13} \text{ m}^2)}{6.7 \times 10^{-11} \text{ kg m s}^{-2} \text{ m}^2 \text{ kg}^{-2}}$$

The units cancel to give:

$$m_2 = 6.0 \times 10^{24} \text{ kg}$$

So the mass of the planet Earth is 6.0×10^{24} kg.

Question 13.2

Since the oil drop is stationary, the net force on it is zero. The gravitational force (F_g acting downwards) can therefore be equated with the electric force (F_e acting upwards). So, using Equations 10.2 and 13.2 gives:

$$mg = Q\mathscr{E}$$

Rearranging to make Q the subject and then putting in the numbers,

$$Q = \frac{6.4 \times 10^{-15} \text{ kg} \times 9.8 \text{ m s}^{-2}}{4.9 \times 10^4 \text{ N C}^{-1}}$$

Since the unit of newtons is equivalent to kg m s^{-2}, the units cancel to give:

$$Q = 12.8 \times 10^{-19} \text{ C}$$

Given that the charge on a proton, $e = 1.6 \times 10^{-19}$ C, this is equivalent to $+8e$.

Question 13.3

Newton's law of gravity was phrased in terms of a force between massive particles. A quantum theory of gravity will describe gravitational interactions in terms of the exchange of gravitons.

Question 14.1

(a) The completed Table 14.1 is shown in Table 14.2.

Table 14.2 Quanta involved in electroweak unification. Completed Table 14.1.

Quanta	Mass energy/ GeV
photon	0
W^+ boson	80
W^- boson	80
Z^0 boson	90
Higgs boson	100–1000

(b) The photon is the only one that is massless. As a consequence of this, Coulomb's law is an inverse square law of force with a very large range. Conversely, W and Z bosons do have mass and so have only a very short range.

Question 14.2

(a) About 12 orders of magnitude span the more-or-less reliable parts of the story; from the 1 eV energy transfers in hydrogen atoms, to the 10^{12} eV (10^3 GeV) investigated at existing or planned particle accelerators.

(b) A further 12 orders of magnitude remain in the range up to a notional scale of 10^{24} eV (10^{15} GeV) conjectured for grand unification.

Question 14.3

(a) Higgs bosons; (b) X bosons; (c) strings and branes.

Question 15.1

If weak interactions did not exist in the hypothetical universe, and if only up quarks, down quarks, electrons, electron neutrinos and their antiparticles were initially present:

- Conversions between protons and neutrons would be impossible.
- As protons cannot convert into neutrons, there would be no proton–proton chain of nuclear fusion in stars.
- As primordial neutrons cannot convert into protons, there would be equal numbers of protons and neutrons in the universe.
- As there are equal numbers of primordial protons and neutrons, these would all combine (eventually) to form nuclei of helium-4. Consequently there would be no hydrogen in the universe.
- As there is no hydrogen in the universe, there would be no water, no organic chemicals, and therefore no life as we know it.

Question 15.2

(a) Primordial nucleosynthesis occurred in the early Universe, between about 100 and 1000 seconds after the big bang. During this epoch, only low-mass nuclei, such as deuterium, helium and lithium, were formed. A second site for nucleosynthesis is in the heart of stars, like the Sun. Here, hydrogen undergoes nuclear fusion to form helium, and later on helium can fuse to form carbon, oxygen, silicon and other (relatively) low-mass nuclei. In fact, most nuclei below an atomic mass of about 62 (nuclei up to iron, cobalt and nickel) can form in the heart of stars in this way. Finally, nucleosynthesis can occur at the end of a star's life, including during a supernova explosion, as a result of the s-process and the r-process. In these processes, many nuclei that are more massive than iron are formed and thrown violently out into the Universe, where they can be incorporated into future generations of stars and planets.

(b) (i) Most of the helium nuclei were formed during the primordial nucleosynthesis, soon after the big bang.

(ii) Most of the oxygen nuclei were formed in the heart of stars.

(iii) All the uranium nuclei were formed as a result of supernova explosions.

Question 15.3

The three key pieces of observational evidence for the hot big bang are:

- the Hubble relationship linking the apparent speed of recession and distance of distant galaxies
- the cosmic microwave background radiation
- the relative abundances of helium, lithium and other light elements.

The first galaxies formed when the Universe was less than a billion years old. So, in theory, observations of distant galaxies only allow cosmologists to reach back this far in time.

The cosmic background radiation last interacted with matter when the Universe was about 300 000 years old. So observations of it only let cosmologists investigate conditions at that epoch.

It is the relative abundances of the light elements that allow cosmologists to reach back the furthest. These elements were formed when the Universe was between 100 and 1000 seconds old, and the reactions that created them were sensitive to things like the ratio of neutrons to protons, which were determined even earlier.

Comments on activities

Activity 3.1

(a–c) The completed version of Table 3.1 is shown in Table 3.2, below.

Table 3.2 Quantum states of hydrogen.

Principal quantum number, n	Orbital quantum number, l				E_n/eV	Total number of quantum states for each n
	0	1	2	3		
1	1s				−13.60	2
2	2s	2p			−3.40	8
3	3s	3p	3d		−1.51	18
4	4s	4p	4d	4f	−0.85	32
Number of quantum states for each l	2	6	10	14		

(d) The fourth energy level comprises 32 quantum states, and in each state, the principal quantum number, $n = 4$. The orbital quantum number is allowed values of $l = 0$, 1, 2, or 3. The magnetic quantum number m_l ranges between $-l$ and $+l$ in each case, and the spin quantum number m_s is allowed values of either $-\frac{1}{2}$ or $+\frac{1}{2}$. Hence the set of four quantum numbers corresponding to each of the 32 quantum states in the fourth energy level is as shown in Table 3.3.

Table 3.3 The set of quantum numbers corresponding to each of the 32 quantum states in the fourth energy level of hydrogen.

n	l	m_l	m_s	n	l	m_l	m_s
4	0	0	$-\frac{1}{2}$	4	0	0	$+\frac{1}{2}$
4	1	−1	$-\frac{1}{2}$	4	1	−1	$+\frac{1}{2}$
4	1	0	$-\frac{1}{2}$	4	1	0	$+\frac{1}{2}$
4	1	+1	$-\frac{1}{2}$	4	1	+1	$+\frac{1}{2}$
4	2	−2	$-\frac{1}{2}$	4	2	−2	$+\frac{1}{2}$
4	2	−1	$-\frac{1}{2}$	4	2	−1	$+\frac{1}{2}$
4	2	0	$-\frac{1}{2}$	4	2	0	$+\frac{1}{2}$
4	2	+1	$-\frac{1}{2}$	4	2	+1	$+\frac{1}{2}$
4	2	+2	$-\frac{1}{2}$	4	2	+2	$+\frac{1}{2}$
4	3	−3	$-\frac{1}{2}$	4	3	−3	$+\frac{1}{2}$
4	3	−2	$-\frac{1}{2}$	4	3	−2	$+\frac{1}{2}$
4	3	−1	$-\frac{1}{2}$	4	3	−1	$+\frac{1}{2}$
4	3	0	$-\frac{1}{2}$	4	3	0	$+\frac{1}{2}$
4	3	+1	$-\frac{1}{2}$	4	3	+1	$+\frac{1}{2}$
4	3	+2	$-\frac{1}{2}$	4	3	+2	$+\frac{1}{2}$
4	3	+3	$-\frac{1}{2}$	4	3	+3	$+\frac{1}{2}$

Activity 5.1

In summary, this activity showed that:

The energy levels of a hydrogen-like ion with nuclear charge Z are more widely spaced than those of hydrogen by a factor Z^2.

The photon energies in the spectrum of a hydrogen-like ion with nuclear charge Z are larger than those of hydrogen by a factor of Z^2.

Activity 7.1

(a) Albert Einstein (amongst many other things) proposed that mass and energy are interchangeable according to $E = mc^2$.

Max Planck first introduced the concept of the quantum into physics as the fundamental amount of energy that an atom can absorb or emit.

Niels Bohr formulated a model of the atom that included electrons in quantised orbits around the atomic nucleus.

Erwin Schrödinger derived a quantum model of the atom in which the positions and velocities of electrons are indeterminate.

Werner Heisenberg formulated the uncertainty principle that rules out the possibility of combining definite knowledge of the position of a particle with definite knowledge of its velocity.

Louis de Broglie proposed that all 'particles' propagate like waves and so will exhibit wave-like behaviour such as diffraction.

Wolfgang Pauli stated the exclusion principle that prohibits any two electrons in the same atom from occupying the same quantum state. [He also first proposed the existence of neutrinos.]

Marie Curie was one of the first people to observe natural radioactivity and isolate the substances responsible for it.

[Paul Dirac has yet to feature in the quantum physics story, but will do so in Chapter 10.]

(b) Obviously, there are no right or wrong answers to this question, and in one sense, virtually everything around you has its basis in quantum physics. Nevertheless, here are the six items that an S104 tutor came up with:

> I see green leaves on a tree outside the window. Their 'greenness' is due to the fact that the chlorophyll molecules in the leaves absorb all colours of light except those photons corresponding to the colour green. This is because the molecules are quantum systems with a set of energy levels that dictate the energies of the photons that they can absorb.
>
> I see the Sun shining in the sky. The light and heat produced by the Sun is a consequence of the nuclear fusion reactions occurring deep within its core, converting hydrogen into helium. These fusion reactions rely on interactions between nuclei that result from the rules of quantum physics.
>
> I have an electric fan on my desk. The motor in the fan is powered by an electric current, which is a flow of electrons. The motion of electrons is governed by the rules of quantum physics.
>
> I sit keying this text at a computer whose internal circuitry is composed of semiconductor devices arranged on silicon chips. These are devices that rely on the rules of quantum physics for their operation.

As I sit here digesting my lunch, chemical reactions are underway in my body metabolising the food I've recently eaten. All chemical reactions rely on the electronic structures of atoms and molecules, which are determined by the underlying rules of quantum physics.

The very fact that I can see things around me is due to the photons emitted by luminous objects, and reflected off other objects, which enter my eyes and are absorbed by the retina, so generating electrical signals that travel along nerves to my brain. All steps of this are quantum processes.

Activity 8.1

As you have seen, the Hubble constant is simply the gradient of the graph of recession speed against distance for a number of clusters of galaxies.
A typical value for the Hubble constant obtained from this activity is around 70 km s^{-1} Mpc^{-1}.

Activity 10.1

Your list will undoubtedly be different from ours, but here are the items we came up with: electric lights, electric toaster, electric kettle, refrigerator, mobile telephone, portable music player, satellite navigation system, computer, electric heater, digital television, and radio.

Of course there are societies today that do not have quite the reliance on electrical gadgets that others do, but it is difficult for most of us to imagine being without at least some of the things in the above list. The world would be a very different place without the technology enabled by Maxwell's insight.

Activity 13.1

(a) The completed Table 13.1 is shown below as Table 13.2.

Table 13.2 Comparing four interactions.

Interaction	Acts on	Particles experiencing	Quanta
strong	colour charge	all quarks (and their antiparticles); gluons	gluons
electromagnetic	electric charge	all electrically charged particles (and their antiparticles)	photons
weak	flavour	all quarks and their antiparticles; All leptons and their antiparticles; W^+, W^-, Z^0	W^+, W^-, Z^0
gravitational	mass energy	All	gravitons?

(b) Some suggested sentences for each heading are show below.

(i) *Quanta*: Strong interactions involve gluons. Electromagnetic interactions involve photons. Weak interactions involve W^+, W^- and Z^0 bosons. Gravitational interactions involve gravitons, though evidence for these is hard to come by.

(ii) *Range*: Electromagnetic and gravitational interactions have a large range, and both forces decrease with the inverse square of distance. Electromagnetic energy is radiated by electric charges that accelerate. This energy propagates through space as an electromagnetic wave. Similarly, gravitational energy is radiated by massive objects that accelerate. This energy propagates through space as a gravitational wave. Both the strong and weak interactions have a very small range, comparable to the size of individual nuclei.

(iii) *Theories*: Strong, electromagnetic and weak interactions are well described by relativistic quantum theories. The first to be developed was quantum electrodynamics (QED), for electromagnetic interactions. This involved combining quantum physics with special relativity and the inclusion of phenomena such as transient electron–positron pairs. Quantum chromodynamics (QCD) describes the strong interaction in a similar way, with the key difference that the exchanged quanta, called gluons, interact with themselves, as well as with quarks. This results in the permanent confinement of quarks and gluons within hadrons, and means that quarks and gluons have never been seen in isolation. The theory of the weak interactions also involves mutually interacting quanta, W and Z bosons. Gravity awaits unification with quantum physics. This makes it hard to get good data on the interplay of gravity and quantum physics. It is also hard to get good ideas about what such a theory would entail, since it must somehow incorporate a 'fuzziness' of space and time.

(iv) *Participants*: Only quarks participate in strong interactions. Quarks and charged leptons participate in electromagnetic interactions. All quarks and all leptons participate in gravitational and weak interactions.

(v) *Strength*: The strong, electromagnetic and weak interactions have strengths that differ only by a factor of ten at energies of 100 GeV. As befits its name, the strong interaction is stronger than the other two. The weak interaction is very weak at low energies, where there is a big price to pay for exchanging its massive quanta. However, at energies around 100 GeV, the comparability of strength with electromagnetic interactions becomes apparent. Again gravity stands out on a limb, due to its extreme weakness at the level of individual particles.

Activity 15.1

(a) Perhaps the simplest way to decide in which order a sequence of events occurred is to think about the energy required for each process. If the processes are then arranged in descending order of energy, they will automatically be in a time-ordered sequence.

Clearly, the formation of the fundamental constituents of matter, quarks and leptons, require the most energy of these processes. This event must have occurred first. Next, as the energy dropped, quarks and antiquarks would have mutually annihilated, leaving behind the relatively few residual matter particles from which to construct the material content of the Universe. Protons and neutrons form next, from the residual quarks. When neutrinos cease to interact with matter, the equilibrium conversion between protons and neutrons effectively stops. After this, the electrons and positrons mutually annihilate leaving relatively few electrons to balance the charge of the protons. From this point on, light nuclei are able to form from the protons and neutrons available. Atoms form next from the nuclei and electrons that now constitute the matter content of the Universe. Finally, background photons interact for the last time with matter when the Universe is about 300 000 years old.

The sequence of the processes listed in the question is therefore:

(iii) formation of quarks and leptons

(vi) annihilation of quarks and antiquarks

(iv) formation of protons and neutrons

(vii) neutrinos cease to interact with matter or radiation

(v) annihilation of electrons and positrons

(ii) formation of light nuclei

(i) formation of atoms

(viii) background photons cease to interact with matter.

(b) A summary of the contents of the Universe at the times indicated is shown in Table 15.1.

Table 15.1 The contents of the Universe at various times.

Section	Time/s	Contents of the Universe
15.2 The very early Universe	10^{-36}	Six flavours of quark, six flavours of lepton, X bosons, photons.
15.3 Inflation	10^{-32}	The same as above, except that X bosons had largely disappeared.
15.4 The quark–lepton era	10^{-5}	Up and down quarks, electrons, positrons, neutrinos and antineutrinos, photons.
15.5 The hadron era	100	Protons, neutrons, electrons, neutrinos and antineutrinos, photons.
15.6 Primordial nucleosynthesis	1000	Mainly hydrogen and helium-4 nuclei; traces of deuterium, helium-3, and lithium 7; electrons, neutrinos and antineutrinos, photons.
15.7 Structure in the Universe	today	Galaxies, stars, gas and dust (all of which are made of atoms, the vast majority of which are hydrogen and helium); photons (cosmic microwave background) fill all space; neutrinos and antineutrinos still present but almost undetectable.

Acknowledgements

The S104 Course Team gratefully acknowledges the contributions of the S103 *Discovering science* course team and of its predecessors.

Grateful acknowledgement is made to the following sources for permission to reproduce material in this book.

Figures

Cover: Eric Heller/Science Photo Library;

Figures 1.1 top centre, 3.9, and 5.2a: Lucy Tindle; Figure 1.1 top right: STSci/NASA; Figure 1.1 bottom left: NASA, ESA and the Hubble Heritage Team (STSci/AURA); Figure 1.2: Courtesy of W.M. Keck Observatory; Figures 1.3, 7.3 and 11.2: CERN Geneva; Figure 3.1: Science Museum/Science & Society Picture Library; Figure 3.8a: Les Chatfield/Flickr Photo Sharing; Figure 3.8b: G W Mullis/istockphoto; Figure 5.2b: Brian Hession/Flickr Photo Sharing; Figure 6.2: Francois Gauthier-Lafaye; Figure 7.1: © Sudbury Neutrino Observatory (SNO); Figures 8.3 and 15.5: NASA; Figure 8.6: Emilio Segre Visual Archives/American Institute of Physics/Science Photo Library; Figure 9.2: Corporate Archives/Lucent Technologies/Bell Laboratories; Figure 10.1: Hulton Archive/Getty Images; Figure 13.1: Mary Evans Picture Library; Figure 13.2: Albert Einstein™ The Hebrew University of Jerusalem, Represented by The Roger Richman Agency, Inc., www.albert-einstein.net; Figure 13.5a: Myers, S. T. et al. 'Quadruple lens system found in the gravitational lens survey', *The Astrophysical Journal*, vol. 447, July 1995; Figure 13.5b Impey, C. D. et al. 'Hubble Space Telescope observations of the gravitational lens system B1422 +231', *The Astrophysical Journal*, vol. 462, May 1996; Figure 13.5c: Wesley, N. et al. 'Unlensing multiple arcs in 0024+1654: reconstruction of the source image', *The Astrophysical Journal*, vol. 461, April 1996, The American Astrophysical Society; Figure 13.6: European Space Agency; Figure 15.4: NASA/WMAP Science Team; Figure 16.1: Mackintosh, R. et al. (2001) *Nucleus, A Trip into the Heart of Matter*, Canopus Books Limited.

Index

Entries and page numbers in **bold type** refer to key words that are printed in **bold** in the text and that are defined in the glossary. Where the page number is given in *italics*, the index information is carried mainly or wholly in an illustration or table.

G

H

I

J

K

L